ENGINEERING SAMPLES IN RACE CARS

It is the first book in a large and special series of books, dedicated to motorsport in general; it will cover aerodynamics, suspension, engines, dynamics, etc. ..... Everything you need to learn how to design a full car.

The aim of this series is also to say that I would like to teach again in a university.

I hope that this series will be a success and that I will be able to transmit all my knowledge and all my experience.

@TimoteoBriet

Design Examples and Systems

The objective of this section is to discuss different systems seen in different seasons and methods, now that we've seen the theoretical principles that govern each part of the car; technically we analyze the adoption of certain paths to always achieve the same objectives: to increase downforce and reduce drag.

DOUBLE DIFFUSER

We have heard of the double diffuser or blown diffuser, methods responsible for making the diffuser act and function optimally, helping the floor of the car. This is take in air and channel it toward the top and bottom of the diffuser:
Top:

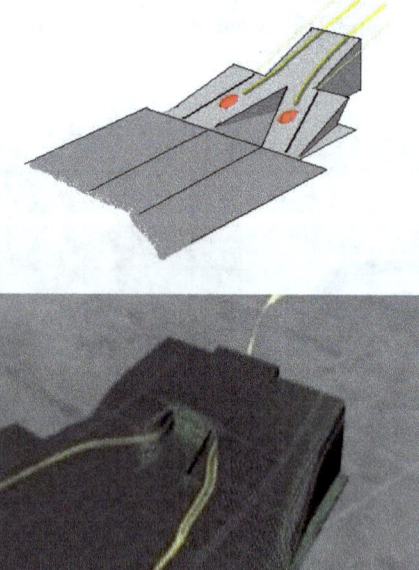

Bottom:

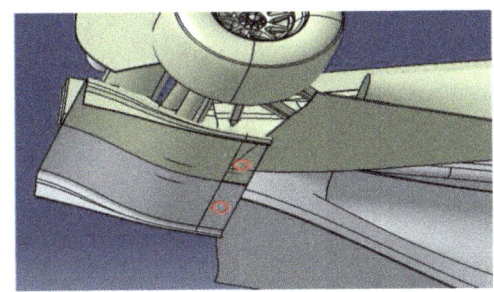

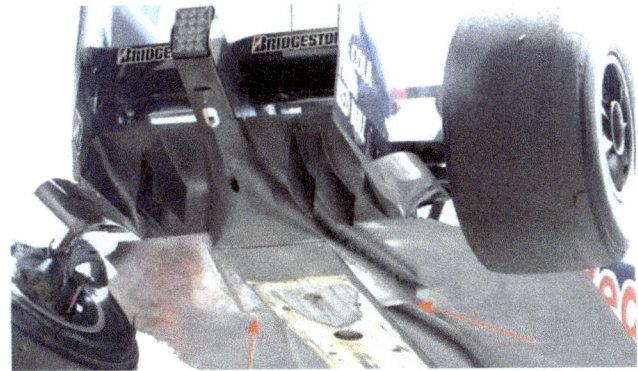

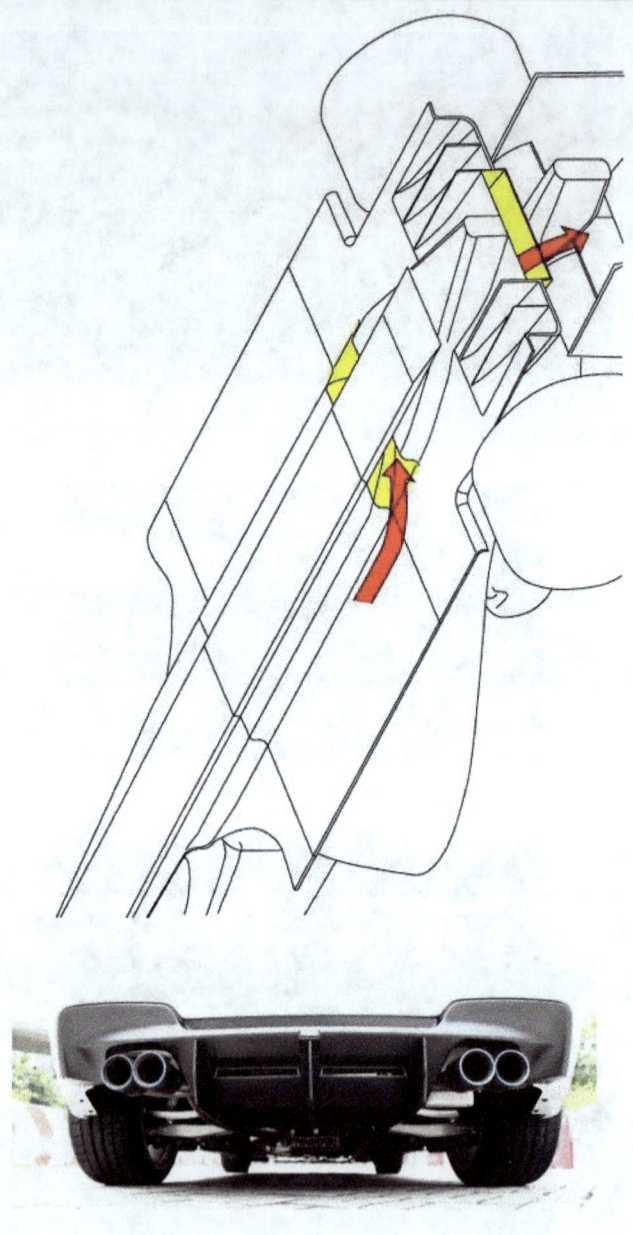

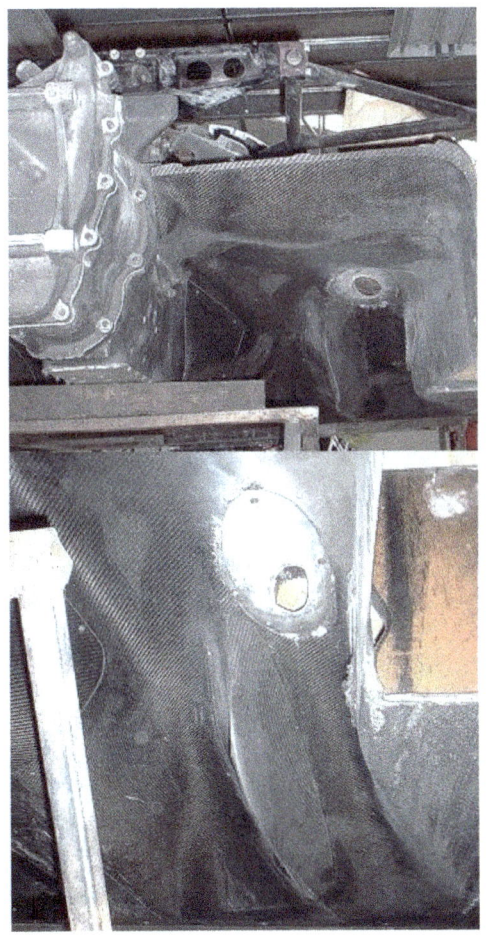

The system is simple in concept; sucks air from "other" place, channeling it toward the top of the diffuser at high speed; this high speed flow of the top reduces the pressure in said zone sucking flow passing below the diffuser, helping the diffuser to work. Thus, the diffuser:

- It can achieve higher angles of incidence.
- But also with smaller incident angle works equally and has less drag, whereby the top speed of the car increases.

In some cars a Gurney flap is placed at the end of the diffuser, with the same aim of producing a low pressure. The Gurney flap is used in many places, provided that an area of low pressure is required.

Now, a little article about, of Willem Toet:

In 2009 the Brawn team won the FiA Formula 1 World Championship. They had a controversial device dubbed a double diffuser fitted to their cars. For years, I was convinced that allowing it was all a conspiracy, both to damage the big manufacturer teams and to help some vulnerable teams step forward, but I now realize that probably wasn't the case.

To understand the various sides of this requires a bit of digging and a little technical understanding as well, but it is the political side and the ramifications the decision had that are most fascinating – you could say that this single interpretation changed the face of Formula 1. Perhaps that's why it is still being dragged up again from time to time today – it has left some people very sore indeed.   One real political twist in the tail, particularly for Honda, was that they gifted the idea for the double diffuser, and hence a chance at the world championship, to what would be the Brawn team. They then did the right thing by the many people in the team by paying a huge amount for the new team's 2009 campaign.   So, Honda had virtually all the financial pain of competing, with none of the glory that passed to the "new" team.

Any of you from inside F1 with views from other teams or just with different perspectives – please let me know what you think. My aim is to use feedback from other inside experts to add to the depth of this review.   I also don't mind including conflicting views if the source and text can be quoted. I'm also willing to consider your views if you feel you cannot officially "go public". I know how hard that is while you're inside the sport. Questions and comments from everyone else also welcome, of course.

Background.

For 2009 there was a substantial change to the regulations. For example: the introduction of KERS, engines could no longer be replaced without limits, bodywork was changed dramatically. All of this meant a reduction in aerodynamic forces and had teams scrambling for any tricks they could find to increase downforce. The double diffuser was a clever interpretation of the rules which gave more of that extra downforce efficiently. Many teams felt (and still do) that the interpretation should never have been accepted and that the real solutions raced were actually illegal.

After a lot of pre-season discussion failed to resolve the issue, double diffuser cars were protested after the first two races. These protests were rejected, first by the Stewards and then at the appeal hearings.

This is a bit of a long one so I've got some headings here to show you what there is....

Rules and interpretation of rules.

Formula 1 has technical and sporting regulations which are accessible to the public. However, this is only part of the story. The FiA issue "clarifications" or Technical Directives which explain how they will interpret the rules as well as private teams correspondence with the same objective. These are expressed as opinions but they are generally considered to be part of the regulations and are also used by the Stewards of race meetings and by appeal court judges as part of their suite of decision making information. Going against these interpretations of the regulations is rarely successful.

The rules and the conventional interpretation of them.

From the middle of 1994 a "plank" was introduced under the floor to stop the cars being run "on the ground" and in 1995 a stepped "parallel plane" floor was introduced (to try to reduce downforce) and make sure air was always passing under the cars. Bodywork in the middle of the car between the rear of the front wheels and the front or center of the rear wheels (rules have changed over the years) has to be "shadowed" by a flat section of floor. The two floor levels are 50mm apart – the central 300mm to 500mm (largely team choice) of the car is shadowed on the lower reference plane. Outboard of that the car is shadowed on the raised "step" plane. If there is no bodywork to shadow there is no need to put floor there. A radius tangential to both surfaces of up to 25mm may be used to blend the two.

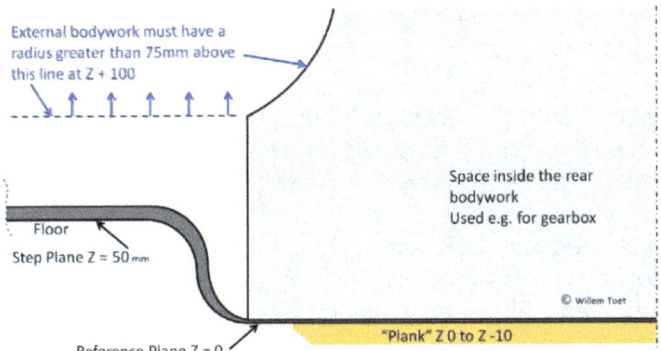

The conventional view. A cross section through the car forward of the start of the diffuser. The optional radius sweeps between the two levels of flat floor.

The rules and the liberal interpretation of them.

The loopholes used to allow the double diffusers were the facts that, if there was nothing to shadow, you didn't need a floor and the fact that you didn't need to shadow suspension. The interpretation used then circumvented the regulation stating that the floor had to be "impervious" (the word used in the regulations). Ross Brawn had obtained written approval for the double diffuser loophole idea.

However, he was cautious about designing his whole car around the idea and came to one of the regular Technical Working Group (TWG) meetings with his aero expert to try to clarify the situation without giving the game away to the other teams. He offered a change to the regulations which would close the loophole. Most teams were completely in the dark and were suspicious of a late change in regulation thinking Ross had a loophole with the new wording rather than the old. The double bluff worked. I was in the meeting representing BMW Sauber. To be approved, the new wording proposed would have required unanimous agreement.

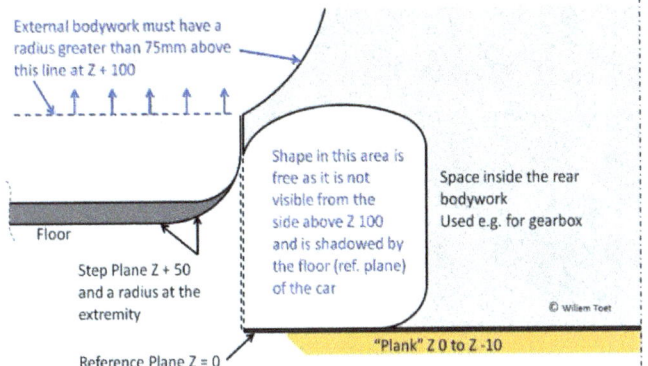

External bodywork must have a radius greater than 75mm above this line at Z + 100

Shape in this area is free as it is not visible from the side above Z 100 and is shadowed by the floor (ref. plane) of the car

Space inside the rear bodywork Used e.g. for gearbox

© Willem Toet

Floor

Step Plane Z + 50 and a radius at the extremity

Reference Plane Z = 0

"Plank" Z 0 to Z -10

This is a cross section through the car, ahead of the diffuser showing one of the ways a double diffuser was given an air entry. It is clear how dramatically different this is to the more conventional interpretation of the rules. Not only is an air entry opened but the size of it is increased by the radius that is used outboard of the slot (this is a regulatory consequence of the interpretation allowing the slot).

There must be no bodywork anywhere directly above the void between the reference and step planes and there must be a void. If the distance in plan view between the two is zero, the two must be jointed together and be impervious.

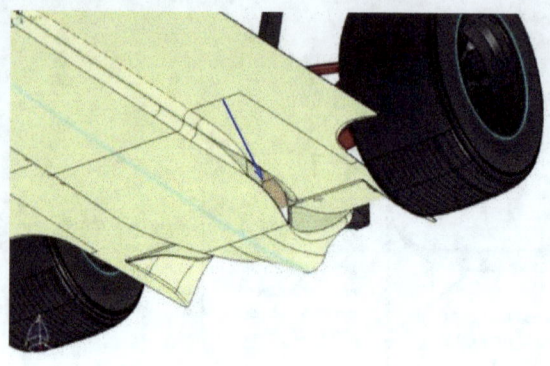

One design example. This is an isometric view from below the car looking at the floor with the (arrow) opening for the double diffuser starting behind the transition between the reference and step planes. There were many ways to do this and the solution chosen was different for each team.

All, though, had an entry forward of the start of the conventional diffuser and an exit linked to a low pressure zone of the car (the rear of the car or under the lower rear wing or both).

The realities of making the double diffuser work

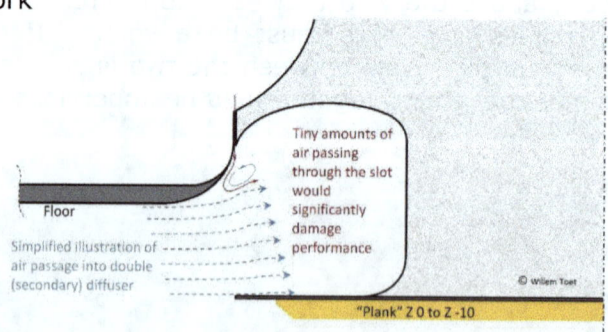

Floor

Tiny amounts of air passing through the slot would significantly damage performance

Simplified illustration of air passage into double (secondary) diffuser

© Willem Toet

"Plank" Z 0 to Z -10

But the design and aerodynamic challenges are just beginning. Finding the optimum shape for the entry and exit for the double diffuser was compromised by the need for the legalizing slot as well as just finding space – especially for those teams that had already designed their cars around a standard diffuser.

At first we considered using a slot 5mm wide (that was the tolerance accepted by the regulations at the time) by the length of the entry for the double part of the diffuser.
That clearly wasn't going to do anything at all positive for the flow which, however you design the system, is operating in an area of very low static pressure and in an area where the air will detach from the surfaces if given half a chance.
The legalizing slot looked like it might make the whole concept useless.

However, if we took advantage of the difference in pressure above the slot (positive pressure), the negative pressure underneath, a really tiny slot and flexible materials, combined with some surface shaping inside the tiny slot we could effectively close it once the car was rolling at low speed.
You could not see through the slot because it was less than 0.5mm wide on the official design and in reality was smaller.

With a bit of force it was possible (flexible materials are wonderful!) to force a small steel rule through the slot to "prove" it was "open".

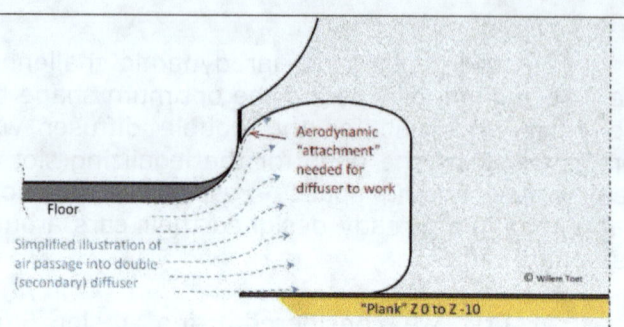

Aerodynamic "attachment" needed for diffuser to work

Floor

Simplified illustration of air passage into double (secondary) diffuser

© Willem Toet

"Plank" Z 0 to Z -10

Note when looking at these illustrative airflow lines that the air mainly moves from the front of the car towards the rear so the flow angularities involved are nowhere near as extreme as they look in this cross section.

Now, honestly the airflow inside the double diffusers wasn't as simple as I have sketched it here, but to show fully realistic flow patterns (these are influenced by height, pitch angle, aerodynamic yaw etc,) would not help the conceptual understanding of the devices. These were (illegal now in F1) interesting things to investigate. The airflow is a little like a half way house between the rotating flow of a (by now) normal motorsport diffuser and the simpler flow of a ducted diffuser.

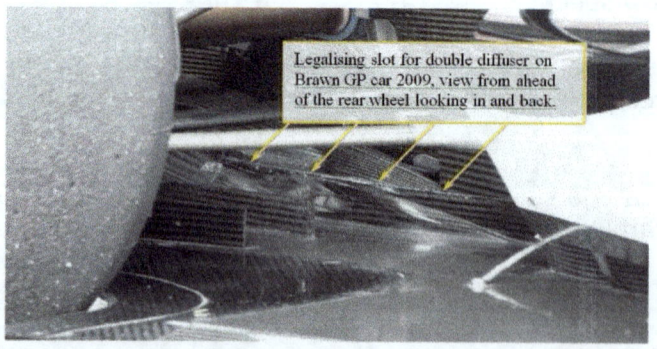

Legalising slot for double diffuser on Brawn GP car 2009, view from ahead of the rear wheel looking in and back.

This picture shows evidence of a tape over back to back. The white marks on the slot surround show the adhesive from race (duct) tape left behind when the tape was removed. I believe that the aero people and race engineers were checking that the sealing mechanism was working correctly.

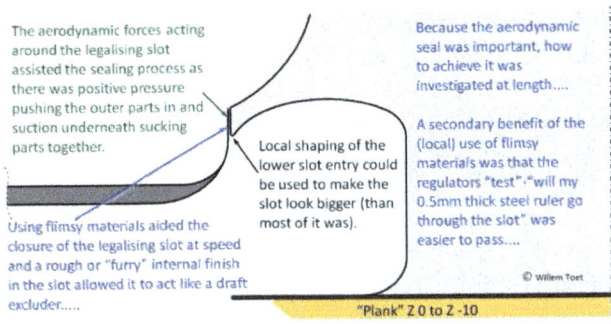

The aerodynamic forces acting around the legalising slot assisted the sealing process as there was positive pressure pushing the outer parts in and suction underneath sucking parts together.

Using flimsy materials aided the closure of the legalising slot at speed and a rough or "furry" internal finish in the slot allowed it to act like a draft excluder.....

Local shaping of the lower slot entry could be used to make the slot look bigger (than most of it was).

Because the aerodynamic seal was important, how to achieve it was investigated at length....

A secondary benefit of the (local) use of flimsy materials was that the regulators "test"-"will my 0.5mm thick steel ruler go through the slot" was easier to pass....

© Willem Toet

"Plank" Z 0 to Z -10

Out on the race track the versions of the double diffuser I know a lot about would certainly have been illegal once the car was up to speed as the slot was closed.
Why the conspiracy theory?

More than 10 years before this loophole was exploited the same regulatory loophole had been used by a small team in another area of the car. Slots and radii were introduced into the floor of the car with the slots mirrored in the bodywork upper surfaces (so you didn't fall foul of the floor's "shadow rule"). It was not allowed because Ferrari (I was there at the time and responsible for asking the polite question to my boss who asked the FiA) pointed out where this loophole would lead. Most of the time you need to take note of historical interpretations but sometimes it is a disadvantage to remember these old interpretations and assume they are still valid – this was one of those rare occasions. Today of course correspondence is shared electronically and is easier to distribute – so questions about new interpretations that don't correspond with an older one are more likely.

We have also all learned to ask in writing – verbal responses are easily forgotten or denied. I had that personal bias in a way that I remembered the much earlier interpretation and wrongly assumed that it would also be remembered by the rule makers.

Adding to the feelings of a conspiracy was that there was a bit of a "war" going on between the rule makers of the time and the major manufacturers. The teams, led by the big players, set up the Formula 1 Teams Association (FOTA) and used this to counter a big push from Max Mosley of the FiA to introduce a budget cap. Alternative, and in my opinion completely fictitious, cost saving ideas were put forward and adopted. Ultimately the teams succeeded in countering the budget cap idea – but totally failed to contain costs. Example – aero testing restrictions. Introduced to cut spend on aerodynamic research. Have failed in part because the rules don't prevent re-allocation of spending and in part because engineers always have counter strategies. It may have reduced the subsequent escalation in spend on aero. For those who would say "no, no, you're wrong, the restrictions have worked" – just look at the size of aerodynamics groups in Formula 1 today. They are huge and have been growing in all the medium-sized teams during the aero testing restriction period.

Who was correct about the best way forward for the future of F1? Well, that depends on your perspective. The biggest, manufacturer-backed teams will almost always win if resources are not restricted because they have the spending power to blow the privateer teams off the race track if and when they choose to do so. It takes some time but the big spending usually works. Sometimes apparently "unlimited" resources are a disadvantage and too much money breeds inefficiency, but a well-organized and efficient large team will always beat a good small one.

The FiA's approach would have meant closer racing and stronger privateer teams. The big teams' approach has allowed them to maintain their advantage – and hence some have stayed in F1. The price has been poorer racing and small teams on the brink of bankruptcy. I am not a politician and have focused on the technical side but you could see that a very different F1 may have been stronger today without some of the big players.

Some literal interpretations of the rules which do, cleverly, comply with the letter of the law are rejected. Teams tried lots of ways to get more air to the diffusers, especially for 2009. For example the curves joining the step and reference planes had to be tangential to the surfaces where they met but nothing in the letter of the law said they had to be tangential in the same direction. So we asked for, and had approved, the following concept :-

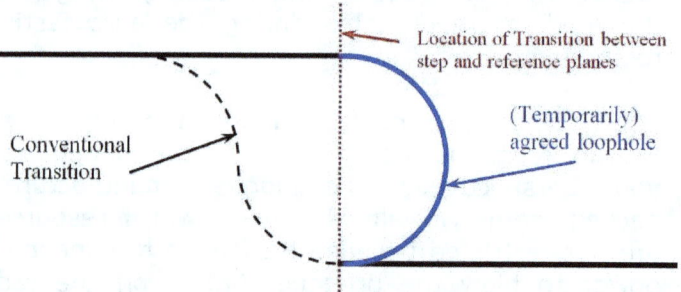

Location of Transition between step and reference planes

(Temporarily) agreed loophole

Conventional Transition

Unfortunately for us, despite having approval in writing from the FiA during 2008 for 2009 and working to that, approval was withdrawn early in 2009. The truth is that it would have been difficult to manufacture – but so was the legalizing slot on the double diffuser. This was one of the reasons we felt there was a conspiracy – that change of mind. We felt, quite strongly, that our interpretation was in complete agreement with the letter of the law and therefore was legal.

How was it that 3 teams started with double diffusers in 2009?

The loophole that was used had been in the rules for years. How could it be that 3 teams miraculously and coincidentally came up with the same concept all of a sudden in 2009. Despite protestations of the time, it wasn't a coincidence.

Honda had announced their withdrawal from Formula 1 and many people, fearing the worst, left the team looking for new opportunities. The normal notice periods for transfer had been waived to save the team money. You can't take designs of listed parts from one team to another but you don't normally unlearn something you've learned! So, this came about due to personnel movement. I will not name names and I don't think anything illegal was done – knowledge transfer is one of the reasons teams poach people and it is one of the reasons that experienced personnel are valuable to others.

Why did Brawn have the best overall double diffuser package initially? There are many reasons but the main one is that they had more development time on it than any other team. Once a person leaves and starts at another team development has to start from scratch. By the end of the season the Brawn team were no longer the fastest but their early point scoring took them over the line.

Politics

The diffusers were permitted in 2009 with the technical people in the FiA focusing on the concept of the interpretation. Having gone through Stewards and appeals it would have been embarrassing to change your mind. Having had the interpretation slapped in our faces the unsuccessful teams had no qualms in taking the interpretation to the limit. By the time rule makers representatives realized we were all "taking the piss" (laughing at the letter of the law in this case) and that actually it had been a bad idea, the rules could not be changed for 2010 (needs unanimous agreement after a certain time the year before). The rules were changed for 2011 with the specification of a large area of the floor which could contain no fully enclosed holes even where there was nothing to shadow. This then forced our devious minds to work on other potential loopholes......

There is another view one could consider. By the time the first races were underway and the protests had been lodged it was clear that the Brawn team had an advantage. Now with the best will in the world it must have seemed quite funny, and good for the sport, to the rule makers that their regulatory interpretation was causing such an upheaval. Having once formed an opinion rule makers like everyone else are human and don't want their opinions challenged – especially not publicly. The teams with the double diffusers were all teams "in trouble". Toyota was threatening to leave (and did in the end). Brawn had zero sponsorship and would not exist in the future if they were not reasonably successful – there had been no buyer for the team remember – only a management buyout and Honda money got them to the grid. And Williams had a dreadful year in 2008 and were looking dodgy. Hence, even if from an engineering perspective you might be tempted to change your mind, there were also good reasons why you should not.

Now we are into opinions, so this is an interpretation.... Why did the Stewards side with the FiA? Well, they are independent (and occasionally show it), but where do you think they get their expertise from? Few really have engineering expertise at the necessary level.

They are likely to ask the FiA technical people for an opinion and for background information. Of course they read the rules but they are a minefield and help is needed. Why was the appeal lost – doesn't bode well when you go against your technical delegate and stewards does it? – not a great precedent to set.   The members of the appeal council also have to be briefed on the details of the technical regulations – and where do they go – same source.   Of course they also read the rules for themselves as well and try to make a sensible call. Once the protest was lost at the race track it was lost and the other teams should have moved on.

Championship position vs. Year for constructors:

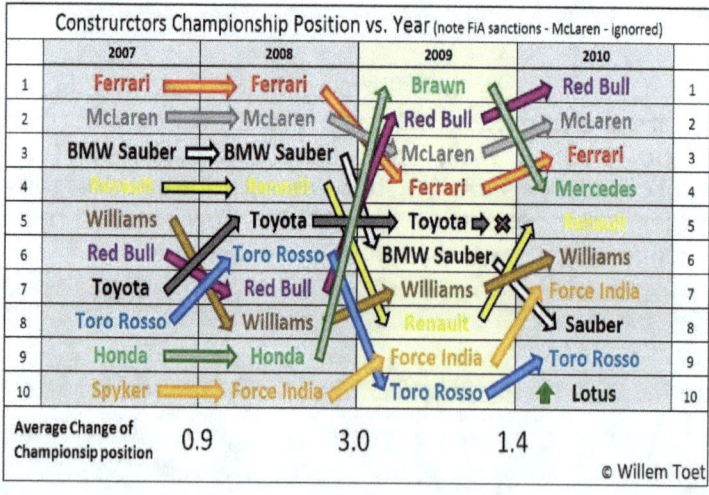

In a year when the rules are relatively stable the average "championship position" change from year to year is somewhere near 1. For 2009 it was 3 and there were two big winners and several losers.

It is clear, from where Honda had been before, that the Brawn team took a huge step forward and that team immediately fell back a little the following year.

They had benefitted from a change to the strongest engine and early commitment to the double diffuser concept. However, parts of the team had been torn apart by the withdrawal of Honda, the resulting cut back in staff and finances as well as the uncertainty that followed. It took Mercedes a little while to rebuild the team.

Red Bull also took a step forward despite not starting with a double diffuser. They invested resources quickly after the realization that they were behind, and had been the quickest car behind the Brawn cars anyway. For BMW it was the end of the road and they withdrew at the end of the year.

Why are details like these never published?

Most teams gag the vast majority of employees. Don't talk to the press is one common instruction (and contractual obligation) for F1 "normal" employees. People who want to keep their jobs are scared. I didn't spill the beans at the time because I was still in F1. Partly that meant being very busy (so no time to hone the words) but partly you won't survive long in the sport, never mind one job, if you bring the sport into disrepute. I was already always on the edge of upsetting my bosses so could not afford to go public (I was one whose communications needed to be sanitized before publication). This is true of most people in F1. It's only a few people who can "legally" talk to journalists without a filter.

Amateurs think they can explain technical things like diffusers and front wings, but clearly have no idea what is behind the engineering side. Their well-meaning and often seemingly plausible offerings proliferate and become folklore.

Professional journalists are not often engineers, are not told the truth by the teams. Race engineers rarely know about aerodynamics in detail anyway and they'd rather invent a plausible story than convey the truth because the teams don't want the truth to be told in case it gives away an advantage. Eventually, long-standing journalists may have earned enough respect to be told the truth, but then they are told that they cannot publish that juicy fact…. I've read every piece I can find on double diffusers and haven't yet found one that I think covers the technical side of the subject really well. In part that's because most articles are short but in part because that combination of willingness to expose and availability of technical facts is missing. The fault really lies with the teams who are so obsessed with secrecy they don't allow information to flow into the public domain.

Why they should have been declared illegal from the start.

I wish to use an example that may not, at first, appear related but is, in my opinion. Ferrari had the rules on bodywork tolerance changed after they'd been excluded from a race for an illegal barge board in 1999. It is precisely this tolerance which I wish to use to illustrate my belief. Between the wheels (precise locations have changed over the years) all bodywork has to be shadowed by surfaces on two planes under the car. The idea of this rule is to limit downforce generation. Ferrari fell foul of this rule because the barge boards were not fully shadowed. Ferrari appealed the decision in two conceptual ways. One was that they didn't gain performance from the error and "proving" this involved opening up CFD and wind tunnel testing results to the FiA. The other was that they insisted there had to be a tolerance to the shadow rule because even they, with all their resources could not achieve infinite precision. A tolerance of 5mm in plan view was accepted and put into the regulations for the following years (it has been modified since to a smaller number).

So, given the legalising slot is somewhere between 50 and 100mm above the reference plane and is tiny, the fact that it has to be precisely aligned with the edge of the reference plane all the way along one side of the slot and given that infinite precision is not achievable even for a big team like Ferrari, it was clear that the real world installations of this concept were going to be illegal to the letter of the law. Even in the last few years some teams have claimed not to be able to manufacture things to a tolerance better than 2mm – a fact that doesn't go so well with claiming they could make the double diffusers legal.

With the benefit of hindsight, talking to other insiders and seeing how development was pursued – I'm now quite convinced none would have passed the letter of the law out on track. Once you see what really had to be done to create double diffusers that actually worked it should be clear that it required infinite precision, a liberal interpretation of tolerances and in most cases, material deflection, to make them work really well. I am willing though, to be educated if someone knows of a legal way to seal the double diffuser legalisation slot.....

DRAG REDUCTION THROUGH THE USE OF WINGS

"We try to minimize drag but
at the same time(if is possible)
increase or maintain downforce."

We know that one of the big problems in competition, if not the biggest, is that if we increase the downforce we increase the drag; a car will be better "aerodynamically" designed if increasing the downforce, the drag does not increase proportionately.

Achieving this is complicated and difficult, but we must try; the basic premises on which we rely are:

- In a straight we need the least downforce and thus less drag in order to achieve higher top speed.
- In turn, we need the greatest possible downforce.

Under these two assumptions or working hypotheses, we will develop three systems to reduce overall drag "only" in a straight.

- Front and rear F-Duct.
- Drag Reduction System or "DRS".
- Automatic F-Duct.

1. Front and rear F-Duct:

Its concept is:

"Take air of a given site, and lead it to another."

Where is the air taken reintroduced? Where harm the downforce, so that the drag and downforce is reduced.
Let's see the rear first:

Rear F-duct:

The car has a number of air inlets via a conduit, it is diverted to the rear spoiler.
To reduce downforce of the rear wing, the system injects air at a given area, usually the crack or gap between the planes binding that make up the entire wing, destroying or damaging the downforce generated. The operation is based on making the air takeoff of the bottom of the wing, allowing to produce downforce:

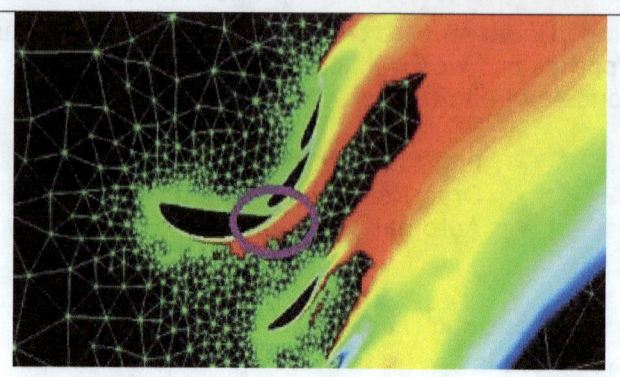

Air injection to "destroy" the downforce, from up or from down:

Where they take the recirculated air to the rear wing? From several parts, depending on the team; McLaren for example, placing the intakes on the nose:

If at any time there is air circulating to the wing, we will "always" have little downforce, which in turn would be very damaging.

For this reason, the tube that connects both areas, is "split". When the pilot reach the straight, connects the two parts, making air to circulate; for this, the pilot cover a hole with his arm or his leg, causing communication:

There may be many variations to this concept, but basically this is the concept.

Front F-duct:

The idea is exactly the same but different application; it is introducing air to "annoy" the downforce on the front wing; thus achieves reduce drag, increasing the top speed.

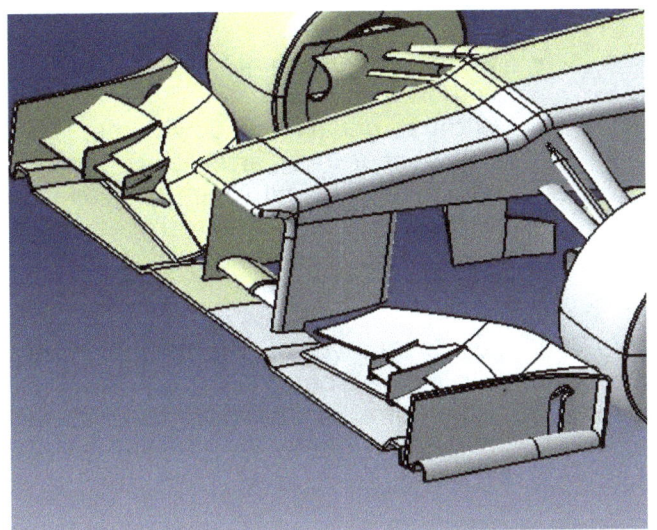

Also air is introduced at the junction of the parts of the front wing.
It can obviously exist and there is a union of two methods: front spoiler and rear wing.

2. DRS:

It is a system analogous to previous but allowed by legislation that is activated at such times and places as the FIA says; this is why the system is much simpler and "clear", it is not necessary "to hide or disguise it" ....

It is a very simple and easy system to understand, in terms of design and in terms of development or way of implementation: reduces downforce of the rear wing when activated. The rear wing has 2 positions:

- Generating downforce (closed or without activating).
- Without generating downforce; To do this separates artificially parts or wings of the wing (open or activated); airflow passes between the gap of the wing without generating load.

It is possible to act on the flap in 2 ways: from the bottom or the front.
Closed or off, open or activated:

With the existence of this device, we have witnessed the manifest superiority of Red Bull, that superiority is based on aerodynamics:

We rely simply on the fact that when the DRS is not activated, i.e. when the rear wing is generating downforce, Red Bull's rear wing incidence is apparently lower than the incidence of other cars on the grid; this entails a further reduction in resistance, and therefore a higher top speed. Red Bull does not need much angle because "has plenty of" downforce.

The spacing between the wings parts are set in regulations (5 cm movement vertical):

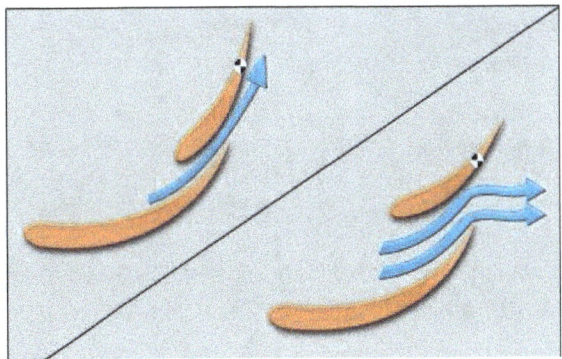

Extrapolating without any regulation to road cars, we can imagine a car with rear wing that suits the driving conditions or meets the conditions that mark the pilot; i.e.: we can do that has little downforce in straights and a lot of it in curves; What's more, we could do that it twist so that in turn could produce more downforce on the outer wheels:

With the activation of the DRS we can also reduce the acceleration time:

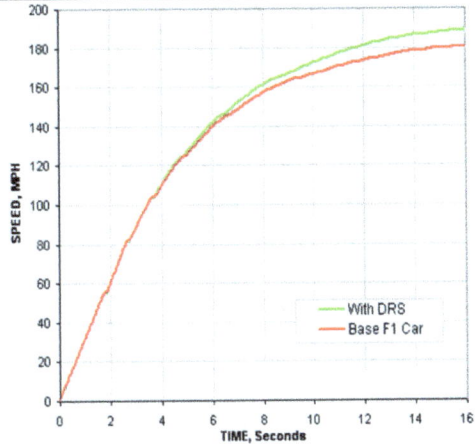

This method is also used in DTM for example:

The "DRS" system not only affects the drag of the car, but also to grip or downforce; see a representative plot of these two effects:

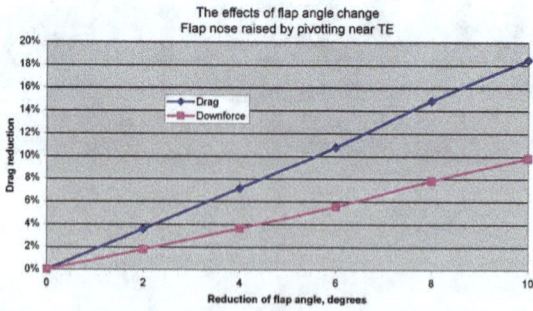

In the 2014 season there were many regulatory changes from a technical point of view; we know too well; This led to a series of structural and geometric changes to the cars, but also in the power plant; it also meant changes in benefits; we will see this change in the new legislation regarding the DRS.

The principle of operation of DRS is based on reduction of drag of the rear wing at some point to thereby, to increase the top speed of the car in order to overtake the car in front.

For this, the upper flap of the rear wing moves, in order to exhibit fewer surface frontally to the air, and make the air flow more easily, reducing the resistance; separation, by law, was 5 cm:

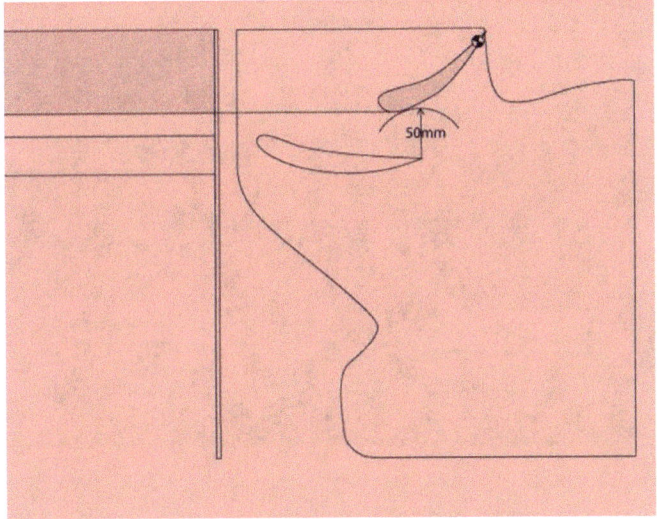

If we make the moving upper flap big, despite moving those 5 cm, remains exposing large surface and producing a lot of drag; so the first objective is to make the smaller better, without losing effectiveness:

The change in legislation in 2014 increased this separation to 6.5 cm, which can make the top flap bigger, and thus reduce drag more:

Thus, as shown in the photo, we can reach the top flap is completely flat even.

Other seasons, Red Bull had the advantage, because the "left over" downforce, could afford to have little upper flap angle without activating the DRS; When activated, the drag reduction was more significant than the other cars.

What we will do now is to quantify this increase in separation, and the equivalent in terms of increased top speed; To do this, we use the CFD simulation: only perform a simulation, incorporating 3 sets of rear spoilers:

1. Wing without activation of the DRS.
2. Wing with activation of DRS – 2013.
3. Wing with activation of DRS – 2014.

Thus, we estimate the drag on each wing, and then calculate the surplus power we will have in our engine in order to achieve higher top speed.
This is the front view of the 3 designs; we can see the increasing gap or opening between parts of the wing:

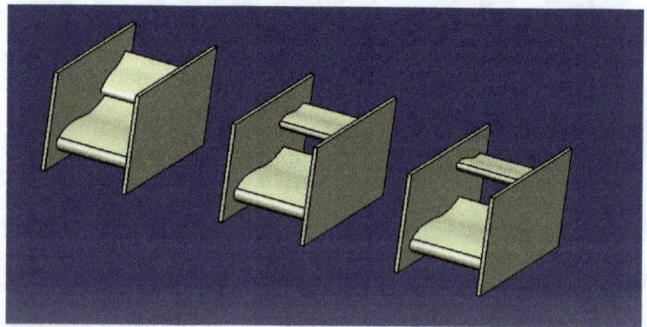

This is a trick that is quite used in CFD simulation, as long as the PC hardware permits it; We are making this simulation on a PC of 16 cores and 48 gigabytes of RAM; still it takes about 28 hours continuously working to finish and get results.... The trick consists of the design of the 3 pieces together, to thus ensure the same conditions simulation to study the pieces.

The results obtained are:

- DRS-2013 activated: 3 % less drag than without DRS.
- DRS-2014 activated : 6 % less drag than without DRS.

On the other hand, we can represent some value fields to see the effect on the airflow that each design and distribution of each design:

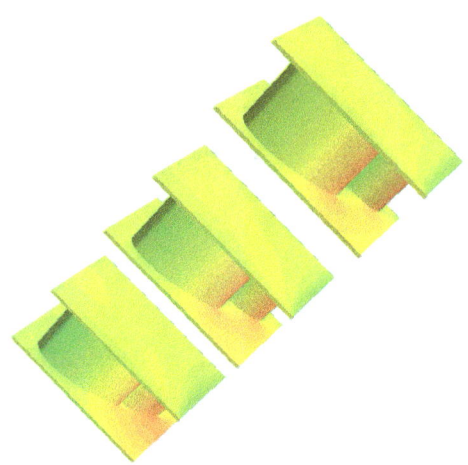

Once obtained drag numerical values, we can see its effect on the top speed of the car; just remember again that this was the aim of the operation of the DRS.

Not to put any mathematical development and assuming certain permissible simplifications, we can assume that if the drag varies to some extent, the speed also varies in the same proportion; this way, we obtain the maximum speeds for each of the three models are:

Assuming that the maximum speed of a F1 is about 310 km/h (by averaging all speed limits on all circuits of the season), we get the maximum speed with DRS in 2013 is about 9 km/h, while in 2014 is 17 km/h. The change has been brutal this season, because practically the speed increase of 2013-2014 has been doubled.

The improvement is evident this season and it "should" allow more overtaking or perhaps provide them.

Finally, we are able to add to DRS "traditional", the effect of reduction drag from rear or front wing; that is:

We connect some holes (free after DRS activation), to front or rear wing in order to "break" the downforce:

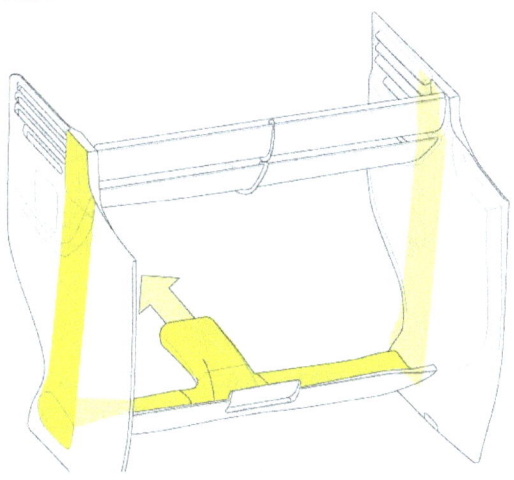

There is another DRS system: the double DRS:

Austrian car has progressed during 2016 to over regularly SF16-H, except Monza. What new technology can be signed autograph Newey in this RB12nbsp; seems that fascinating and unique development of the concept of 'rake' would be one of the reasons.

Technicians McLaren, Red Bull, Ferrari and Co. are struggling to mitigate the differences with Mercedes power through aerodynamics. But Daniel Ricciardo got to reach the 356 km / h top in Monza, about 4 km / h less than the Mercedes with a propeller Renault still behind rivals Honda least. In Germany, the Austrian car also shone especially.

The RB12 would earn top speed with a sort of 'double DRS': the conventional aileron, and formed by the aerodynamic behavior of the chassis through the particular rake developed by Red Bull and Newey. This is explained graphically in 'Motorsport.it' Enrique Scalabroni Argentine engineer, who in his day was part of Williams and Ferrari.

Newey introduced since 2010, the 'rake' is the difference in height above the floor between the front and rear of the car, thus helping to create a 'vacuum effect' with the entire chassis. The diffuser approaches the asphalt as the car gains speed, thus reducing the rake '. The rear wing thus reduced its 'drag' and increases the top speed. In its present application could be a key in the progression of Red Bull in 2016.

"The angle generated by the bottom of the car allowed to have a diffuser with a much larger front to rear exhaust surface," explains Scalabroni. Regarding the rake, "the problem is that, to generate the necessary burden, had to seal the sides to prevent air introducing in floor, which prevents forming a zone of depression." At the time of the 'wing car' that role was covered by a movable flaps on the sides of the flanks. Today they are prohibited.

Newey used from 2010 exhaust blown way 'sealed', which Scalabroni called "thermal skirts". Also prohibited now, the British engineer would have developed what Scalabroni called a "pneumatic skirts". And how these 'virtual' skirts work?

"Timely Harnessing the vortex (air) generated near the neutral area of ??the above ailerons, a flow called by technicians '250' because it is a turbulence that is generated at 250 mm from the center line of the car," Scalabroni explains. Therefore it is sought feed and redirect the flow to the bottom side of the car, in a sort of "air flaps" that seal the bottom (of the car) with asphalt, only at high speed. "

Scalabroni explains that, at first, Red Bull stood back 50 mm, but has reached 150. "In previous years, the cars of Newey had an enviable cornering, but did not shine at maximum speeds. Now, Newey has found the answer to this problem. For the first time, with increasing speed and increase the load, the rear is lowered. " This would be the second DRS that Scalabroni spoke ...

Suspensions, soft as butter:

How the 'fall' of the rear speed is achieved according win? "Riding very soft springs on the suspension. While other teams are bent on turning with springs 1,500 pounds because they want to keep the rake, Red Bull does not exceed 600 pounds. "

To illustrate, Scalabroni go to a recording of Canada "that showed how Ferrari could reduce drag rear spoiler and 'monkey seat' to advance by flex the central support of that, while technicians Red Bull prefer to download all rear, and with some progressivity, which makes the frame more stable and manageable. "

Therefore, according to the Argentine coach, "the Ferrari, in acceleration out of the bend, cannot have the same drive that has a very stiff suspension, while the RB12 absorbs any unevenness of asphalt, ensuring a 'grip' enviable thanks to the softer springs. " complaints of Vettel and Raikkonen well remember the nerve rear of his car ...

Furthermore, to "balance it, incorporated in the Forewing one 'flap' which has a flexibility programmed to compensate for loss of downforce on the rear spoiler, which can maintain a good balance allows also not punish gums during stint" .

A mechanism for straight:

This mechanism works only in the straight and not curved, "because under braking, the chassis front loading with weight transfer, thereby restoring the back height and therefore downforce generated rake". That is, the RB12 would have great intrinsic support and stability in tacking areas, with a system to compensate for the lack of power in straight ...

In short, it would be a brilliant development of a concept, not known, less difficult to optimize. "Achieving a profit of rake is a long and complicated process," said Pat Symonds regard, technical manager of Williams, "you have to develop everything around it, and the gains are incremental. You cannot run with rake and think that you are already going as fast .... " However, it seems that Adrian Newey has again achieved.

3. Automatic F-duct

The problems that owns the F-Duct system apart from other considerations, are legal issues; ie If for legislation is or is not accepted; For this reason, engineers have managed to design an F-Duct which is activated automatically; that is, without driver input.

To do this, they have opted for 2 options:

   a) We know that at a certain speed through a flange, not passes more airflow; ie for a given flange, there is a maximum mass flow rate through the conduit.

   b)

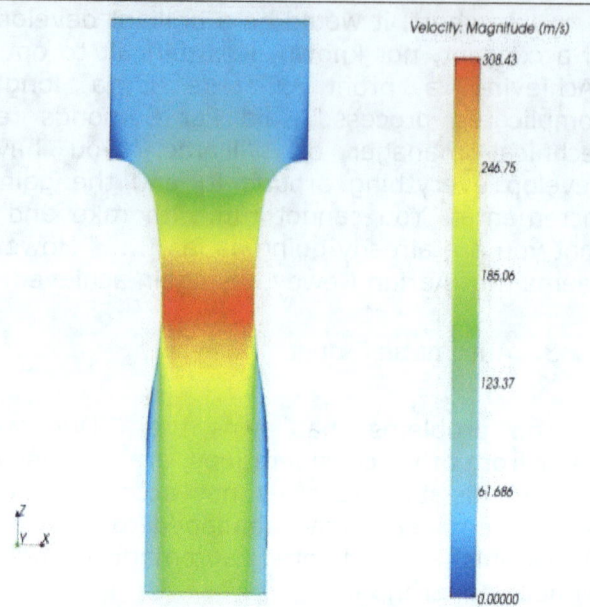

Velocity: Magnitude (m/s)

308.43

246.75

185.06

123.37

61.686

0.00000

c) We identify the plug as the dark central zone; this is called sonic plug, and depending on the dimensions of the conduit occurs at a certain speed the car.

The brilliant idea involves placing a pipe-flange between the two parts of the wing, so that the plug is formed:

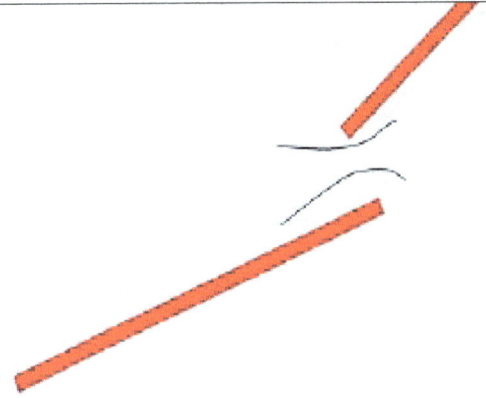

Thus, without any intervention by the driver, at a certain speed, not more airflow passes between the wings of the wing, producing no more down force and therefore more drag; By contrast, curve, speed is less than the maximum line speed, so we have enough downforce.

May be, is possible to create another automatic system, as the last; the function is the same:

Another method in order to reduce the downforce (and so, the drag) is to reduce the gap between wing parts, from deflection material. That, can produce also, the formation of tobera geometry. Is very complicate that....

The other used method exploits the movement of DRS for destroying the flow over the wings: slots are placed on the extreme rear spoiler screens that when the DRS is activated they are exposed and are the beginning of conduits; these ducts carry air to destroy the downforce on the front wing.

It's an elegant way to increase the efficiency of rear DRS, with a front DRS.
In the case of Mercedes, the flow enters the front wing at the front through some cracks:

Mercedes used this system to channel air to the front wing, to generate a loss of front load, plus rear.

But this double DRS has drawbacks; remember the case of Schumacher who had several accidents at the end of straight having activated double DRS;

The problem is double:

- On the one hand, being behind another car, the air he receive is "dirty" and therefore does not have so much downforce, which needs more space to brake; i.e. should brake before.
- Furthermore, having less load in the front train, also you need more braking distance because it has less front grip.

As we see, they are two distances to be combined; ie: Schumacher misjudged the braking distance, but also influenced by these two distances above.
In fact the FIA banned the double DRS in 2013 ....

→ Something could be done, and perhaps by law cannot be:

This is to introduce automatically flow into the floor or in the car diffuser; sure reducing downforce generated by the floor or diffuser. We would get much faster top speed. Remember the great load that the floor-diffuser generates.

Why the DRS in not placed in front wing ???? good question....

Let aero values (drag coeff) for some car with open wheels (13.5 mm front height and 20 mm rear): horizontal (front angle) and vertical (rear angle):

SCX									FRONT FLAPS															
REAR WING	5	6	7	8	9	10	11	12	13	14	15	16	17	18	19	20	21	22	23	24	25	26	27	28
5	0.343	0.343	0.350	0.350	0.351	0.351	0.351	0.352	0.353	0.353	0.353	0.354	0.354	0.354	0.355	0.355	0.355	0.355	0.355	0.355	0.355	0.355	0.355	0.355
6	0.358	0.358	0.358	0.353	0.353	0.360	0.360	0.361	0.361	0.362	0.362	0.362	0.363	0.363	0.363	0.363	0.364	0.364	0.364	0.364	0.364	0.364	0.364	0.364
7	0.368	0.368	0.363	0.363	0.370	0.370	0.370	0.371	0.371	0.372	0.372	0.372	0.373	0.373	0.373	0.374	0.374	0.374	0.374	0.374	0.374	0.374	0.374	0.374
8	0.380	0.380	0.380	0.381	0.381	0.382	0.982	0.982	0.983	0.983	0.984	0.984	0.984	0.385	0.385	0.385	0.385	0.385	0.386	0.386	0.386	0.386	0.386	0.386
9	0.392	0.392	0.393	0.393	0.394	0.394	0.394	0.395	0.395	0.396	0.396	0.936	0.397	0.397	0.397	0.397	0.398	0.398	0.398	0.398	0.398	0.398	0.398	0.398
10	1.005	1.005	1.005	1.006	1.006	1.007	1.007	1.008	1.008	1.008	1.009	1.009	1.009	1.010	1.010	1.010	1.010	1.010	1.011	1.011	1.011	1.011	1.011	1.011
11	1.033	1.034	1.035	1.035	1.035	1.035	1.035	1.035	1.035	1.034	1.034	1.034	1.033	1.033	1.033	1.033	1.033	1.033	1.034	1.034	1.035	1.036	1.037	1.037
12	1.046	1.047	1.047	1.047	1.048	1.048	1.047	1.047	1.047	1.045	1.045	1.045	1.045	1.045	1.045	1.045	1.045	1.045	1.046	1.046	1.047	1.048	1.050	1.050
13	1.058	1.058	1.059	1.059	1.059	1.059	1.059	1.059	1.058	1.059	1.050	1.057	1.057	1.057	1.057	1.057	1.057	1.057	1.058	1.058	1.059	1.060	1.061	1.061
14	1.063	1.063	1.070	1.070	1.070	1.070	1.070	1.070	1.063	1.059	1.058	1.058	1.058	1.058	1.058	1.058	1.068	1.069	1.063	1.070	1.071	1.072		
15	1.073	1.073	1.080	1.080	1.081	1.080	1.080	1.080	1.073	1.073	1.078	1.078	1.079	1.079	1.079	1.079	1.079	1.079	1.080	1.081	1.083			
16	1.088	1.083	1.089	1.090	1.090	1.090	1.090	1.090	1.089	1.088	1.088	1.088	1.087	1.087	1.087	1.088	1.088	1.089	1.090	1.091	1.092			
17	1.097	1.037	1.098	1.098	1.098	1.098	1.098	1.098	1.096	1.096	1.096	1.096	1.096	1.096	1.096	1.096	1.097	1.097	1.098	1.099	1.100			
18	1.104	1.105	1.105	1.106	1.106	1.106	1.106	1.106	1.104	1.104	1.104	1.103	1.103	1.104	1.104	1.104	1.104	1.105	1.105	1.307	1.108			
19	1.111	1.112	1.112	1.113	1.113	1.113	1.112	1.112	1.111	1.111	1.111	1.110	1.110	1.111	1.111	1.111	1.112	1.112	1.113	1.114	1.115			
20	1.118	1.118	1.119	1.119	1.119	1.119	1.119	1.119	1.118	1.117	1.117	1.117	1.117	1.117	1.117	1.118	1.118	1.119	1.120	1.122				
21	1.124	1.125	1.125	1.126	1.126	1.126	1.125	1.125	1.124	1.124	1.124	1.123	1.123	1.123	1.123	1.124	1.124	1.125	1.126	1.127	1.128			
22	1.131	1.131	1.132	1.132	1.132	1.132	1.132	1.132	1.131	1.131	1.130	1.130	1.130	1.130	1.130	1.131	1.131	1.132	1.133	1.134				
23	1.138	1.138	1.139	1.139	1.139	1.139	1.139	1.139	1.138	1.138	1.137	1.137	1.137	1.137	1.138	1.138	1.139	1.140	1.141					
24	1.146	1.146	1.147	1.147	1.147	1.147	1.147	1.147	1.146	1.146	1.145	1.145	1.145	1.145	1.145	1.146	1.147	1.148	1.150					
25	1.155	1.156	1.156	1.157	1.157	1.157	1.157	1.156	1.156	1.155	1.155	1.154	1.155	1.155	1.155	1.156	1.157	1.158	1.159					

SCX									FRONT FLAPS				
REAR WING	5	6	7	8	9	10	11	12	13	14	15	16	17
5	0.343	0.343	0.350	0.350	0.351	0.351	0.351	0.352	0.952	0.953	0.953	0.953	0.954
6	0.358	0.358	0.358	0.353	0.353	0.360	0.360	0.361	0.361	0.361	0.361	0.362	0.362
7	0.368	0.368	0.363	0.363	0.370	0.370	0.370	0.371	0.371	0.372	0.372	0.372	0.373
8	0.380	0.380	0.382	0.381	0.382	0.982	0.982	0.982	0.983	0.983	0.984	0.984	0.984
9	0.392	0.392	0.393	0.393	0.394	0.394	0.394	0.395	0.395	0.396	0.396	0.936	0.397
10	1.005	1.005	1.005	1.006	1.006	1.007	1.007	1.008	1.008	1.008	1.009	1.009	1.009
11	1.033	1.034	1.035	1.035	1.035	1.035	1.035	1.035	1.035	1.034	1.034	1.034	1.033
12	1.046	1.047	1.047	1.047	1.048	1.048	1.047	1.047	1.047	1.047	1.046	1.046	1.046
13	1.058	1.058	1.059	1.059	1.059	1.059	1.059	1.059	1.059	1.058	1.058	1.058	1.057
14	1.063	1.063	1.070	1.070	1.070	1.070	1.070	1.070	1.070	1.070	1.063	1.063	1.063
15	1.073	1.073	1.080	1.080	1.081	1.081	1.080	1.080	1.080	1.080	1.073	1.073	1.073
16	1.088	1.083	1.089	1.090	1.090	1.090	1.090	1.090	1.089	1.089	1.088	1.088	1.088
17	1.097	1.037	1.098	1.098	1.098	1.098	1.098	1.098	1.098	1.097	1.097	1.097	1.096
18	1.104	1.105	1.105	1.106	1.106	1.106	1.106	1.106	1.105	1.105	1.105	1.104	1.104
19	1.111	1.112	1.112	1.113	1.113	1.113	1.113	1.113	1.112	1.112	1.112	1.111	1.111
20	1.118	1.118	1.119	1.119	1.119	1.119	1.119	1.119	1.113	1.113	1.118	1.118	1.118
21	1.124	1.125	1.125	1.126	1.126	1.126	1.126	1.126	1.125	1.125	1.125	1.124	1.124
22	1.131	1.131	1.132	1.132	1.132	1.132	1.132	1.132	1.132	1.132	1.131	1.131	1.131
23	1.138	1.138	1.139	1.139	1.139	1.139	1.139	1.139	1.139	1.139	1.138	1.138	1.138
24	1.146	1.146	1.147	1.147	1.147	1.147	1.147	1.147	1.147	1.147	1.146	1.146	1.146
25	1.155	1.156	1.156	1.157	1.157	1.157	1.157	1.157	1.156	1.156	1.156	1.156	1.155

We can see that if we change ONLY the front angle, the full drag is practically the same.... That is because the air from behind is very POOR, son there will be higher front drag but lower rear drag (the adition, the same). So is not possible to place the DRS system in front wing (also with understeering....).

Other automatic system, in order to reduce drag:

Pressure +0.9 (Cp) Press +1.0 **5** Pressure -1.5 to -0.5

Press +0.0 **3** Pressure -1.2 to -0.7

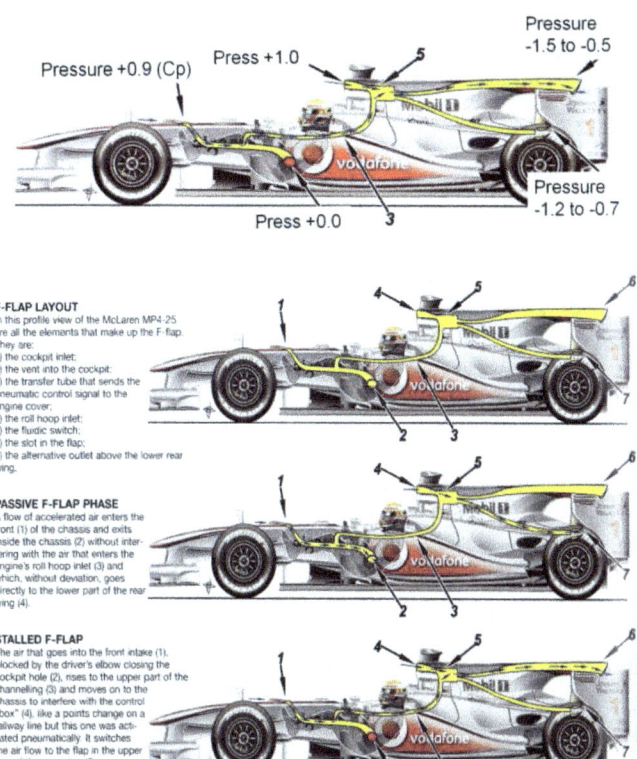

F-FLAP LAYOUT
In this profile view of the McLaren MP4-25 are all the elements that make up the F-flap. They are:
1) the cockpit inlet;
2) the vent into the cockpit;
3) the transfer tube that sends the pneumatic control signal to the engine cover;
4) the roll hoop inlet;
5) the fluidic switch;
6) the slot in the flap;
7) the alternative outlet above the lower rear wing.

PASSIVE F-FLAP PHASE
A flow of accelerated air enters the front (1) of the chassis and exits inside the chassis (2) without interfering with the air that enters the engine's roll hoop inlet (3) and which, without deviation, goes directly to the lower part of the rear wing (4).

STALLED F-FLAP
The air that goes into the front intake (1), blocked by the driver's elbow closing the cockpit hole (2), rises to the upper part of the channelling (3) and moves on to the chassis to interfere with the control "box" (4). Like a points change on a railway line but this one was activated pneumatically. It switches the air flow to the flap in the upper part of the rear wing (5).

This system is very important; in fact, (5) deflect the air automatically, against the speed (for example, 250 km/h).

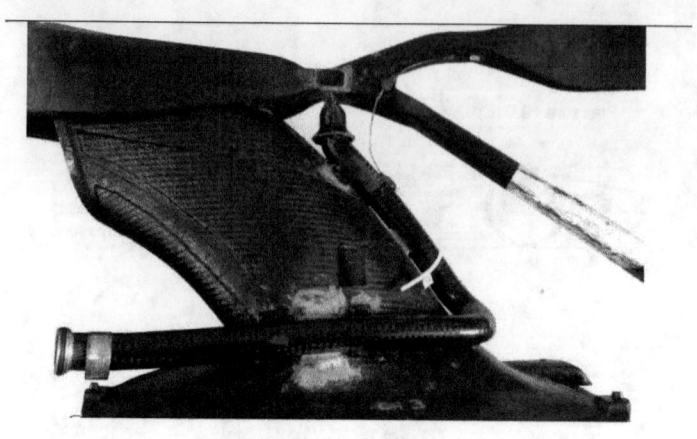

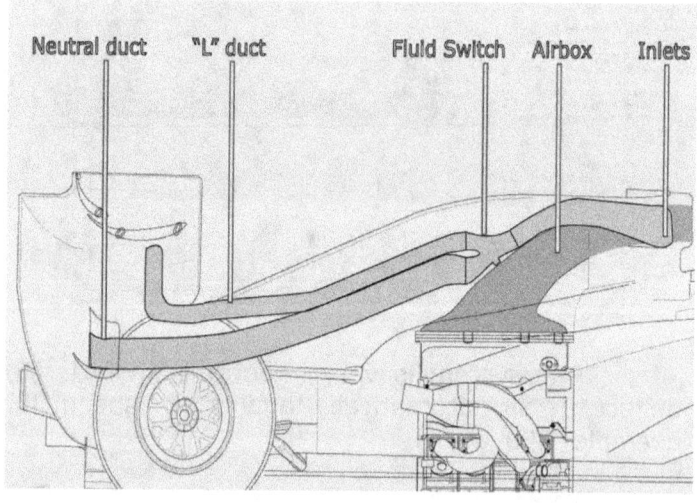

Neutral duct   "L" duct   Fluid Switch   Airbox   Inlets

In the last drawings, we have added some information about the expected general air pressure in the important zones. If you stick your hand out of a moving car you can feel the air pressure acting on it both as high pressure (forward facing parts in the air) and low pressure (you maximise this suction pressure if you present a curved surface facing down a bit like the biggest element in my wing sketches above).

You will notice that you have positive or neutral pressures entering the various ducts and negative pressures at both exits. So the obvious question is, why would airflow simply not permanently exit out of both exits? This is what the inventor of the concept at McLaren and then other teams had to grapple with and this is what has not been pointed out adequetly. The flow switching is done using a combination of ducts which cross over one another (and each team had different duct shapes).

Craig - Scarbs, on his website, has been very smart and not really tried to explain the finer details of the aerodynamics in a theoretical way. He's used his resources to get up close and personal with some actual F-Duct pipework. This one is from Force India as many teams kept their pipework well hidden. This version shows what many teams had to do to get their F-ducts to work – create areas of contraction and a switching "jet" to do the flow "switch" itself. Just one look at this image tells an aerodynamicist that this is not a simple or quick to optimise concept.

The original inventor must have been both persuasive (to convince people that the work needed to be done) and persistent (to develop a working solution).

With this particular F-Duct, the air coming from the upper entries is squeezed into a jet, and, using a slight curve (Coanda effect) flows naturally into the lower duct.

The upper part of the duct has a virtual sharp angle at its top (end of contraction) close to where the cross over occurs so that does not encourage flow towards the following upper duct. The switching duct, where it joins the others almost points forward, and this in turn, discourages air from exiting and in fact air would probably flow into that duct (and the cockpit) if the pressures in the cockpit were identical to those at the rear exit of the lower duct.

When the driver blocks the cockpit exit for the F-duct forward entry, he removes the cockpit exit and creates a sudden change in air pressure in the switching duct. It now has enough pressure to create a small jet which detaches the Coanda effect flow sending air to the lower duct and pushing it up so that the flow from the upper entries can "see" the upper surface of the duct junction. It is significant that the disrupting (switching) duct enters the assembly at a point behind the kink in the upper surface. Coanda effect then attaches the flow to the upper duct's surface and the switch has been achieved.

A complication to this is that when the F-duct stalls the wing, the pressures change – especially for the upper duct and it is possible that the duct could switch back. Aerodynamic hysteresis we'd call it. Happily for mere aero people like us, there was enough change in the switching duct pressure when the driver "sealed" the cockpit exit, that this was able to overcome the likely aerodynamic hysteresis.

Various teams have tried (and even raced) passive versions of fluidic switches using one duct that is fed by turbulent air and another that is fed by laminar flow. We're getting into aerodynamic detail here, but the way energy loss inside a laminar flow duct changes with (vehicle) speed is different to the way losses evolve with turbulent flow. This, though, is a more subtle way to change which duct has more "power" and the feedback hysteresis from the wings to switch back and forth is hard to overcome. I don't know of anyone racing such a system today but I think it would still be possible to find ways to counteract that hysteresis. Slots in the wings themselves are illegal now so different mechanisms have to be found (and had been found). One day when the rules make what I'm thinking illegal I'll spill the beans (as they say).

Note that resources are always limted (even in F1) and the teams have to have a reasonable expectation of success before they invest research capability in a new idea. That's why I believe the person who invented what became known as the F-Duct must have been persuasive and persistent. I've worked with a lot of people like that and they are difficult people. They can bring fantastic steps in performance, but also they can waste lots of time and energy and bring no reward. They are, though, the ones that take the science forward to new levels of understanding.

JORGE LORENZO HELMET; INCREASED TOP SPEED AND REDUCTION OF TURBULENCE

The drivers or teams hide or attempt to hide aerodynamic devices behind some spectacular appearances; Adrian Newey, banned a few seasons ago Webber used a brand of helmet because he could adversely affect the performance of the car.... We analyze, this time, the role of helmets.

All we can intuit the aerodynamic function that can have helmets, both car and motorcycle; Motorcycle helmets may be more important from this point of view, because they are more exposed to air flow and determine further the flow going towards the rear. But we must also bear in mind that F1 drivers' helmets are located very close to the engine intake, with all that that entails of engine power output and its consequences.

Several seasons ago, we were tasked to modify the Jorge Lorenzo helmet when he ran last season in 250 cc; the Spyder used helmet of NZI Helmets; Jorge was complaining about very annoying vibration at the time, of course, to drive; and the task was precisely this: eliminate or at least mitigate them.

To do this, and given the short time between races had, just 10 days, we started a study in 2 dimensions; To do this, we take some side photos of Jorge over his bike in his own box:

Once we had the design in CAD, we performed a CFD simulation to see the status of the problem, obtaining a map of pressure and velocity vectors:

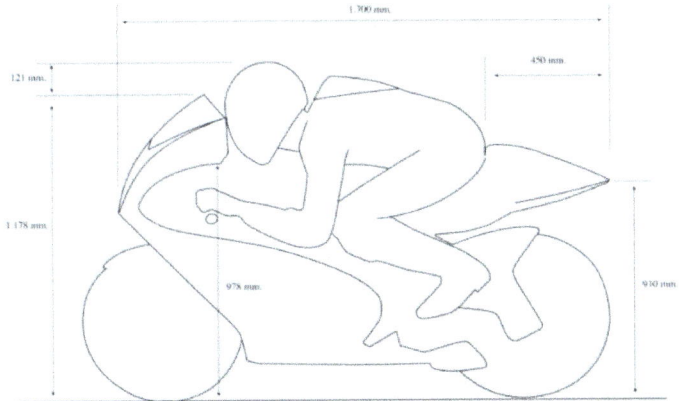

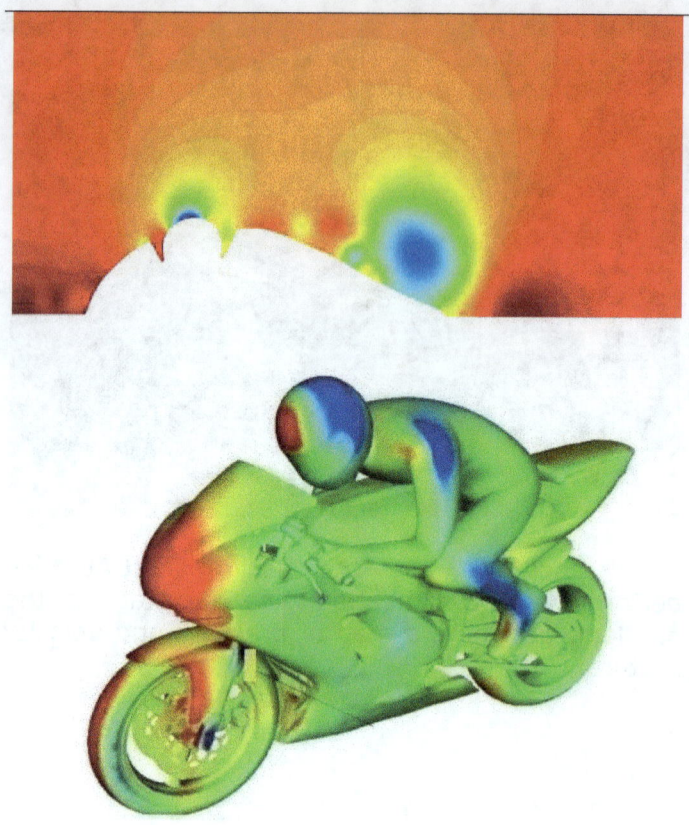

The importance of this first study was to observe that in the back of Jorge there were recirculation and turbulence; this turbulence originated vibrations which Jorge complained:

This recirculation was caused by excessive pressure gradient:

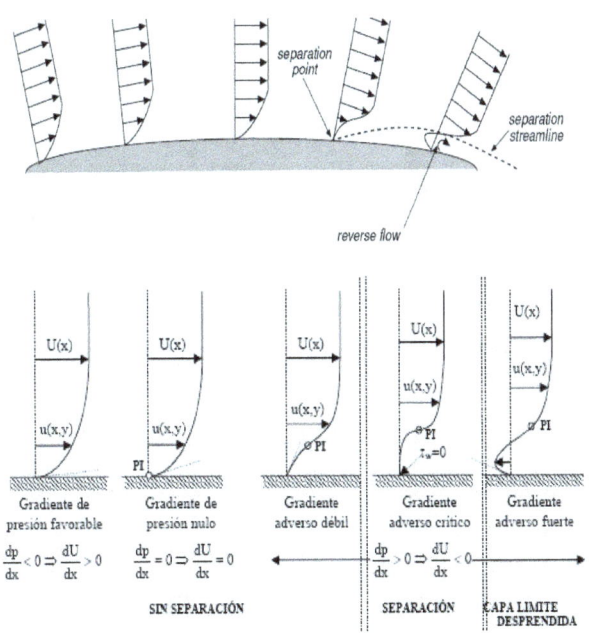

What should we do to eliminate them? Play around with the back of the helmet: for this, we put a kind of appendix:

The incorporation of this appendix immediately involved a reduction or elimination of vibrations on the pilot:

Furthermore, the removal of this turbulence also included the reduction of drag; Later it was learned that the bike had a top speed 5 km / h more.
The helmet in issue was left as follows:

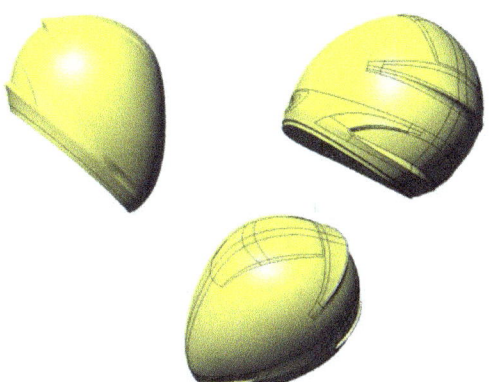

It is a clear example of the importance of helmet in high competition; in the case of cars helmets are different even conceptually, but have two basic functions:

- Not cause excessive drag and turbulence.
- Adjust the airflow towards the higher engine intake.

Just because reducing the aerodynamic drag, it also means that it does not produce turbulence backward, the flow of air to the rear wing is clean and profitable for it.

Most F1 helmets current and past seasons have this little appendix that has the Spyder:

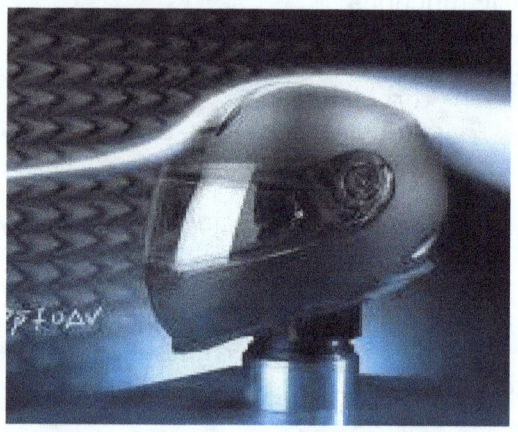

The particular and interesting fact of the design of the appendix, unlike the vast majority of other helmets F1, is connecting the rear bottom with the top; This is called auto stable system. The question is: the more speed the car has, the less pressure there will be on the lower back of the helmet, which said depression sucks air from the top filling said prior depression, reducing resistance. It is called self-supporting, since it is the system itself that is regulated as a function of speed.

By filling this depression does not mean only a reduction in aerodynamic drag and therefore a more efficient rear wing; remember what was did with Jorge Lorenzo: to increase the top speed by 5 km / h, avoid vibrations and prevent dirty air back.

CARAVANS' DRAG RECUCTION

We've all seen this characteristic car and caravan stamp; the drag that caravan produces is aerodynamically huge and therefore in terms of fuel consumption; we can work on two ideas:

- Modify the geometry of the caravan.
- Incorporate "something" to the car to deflect the flow and does not hit directly into the caravan.

The first method is very complicated (regulations and other considerations); for the second case, simply put with a spoiler to separate the flow that is going back. The benefits are enormous, as we can see in the graphs and the system is simple, easily adaptable to any vehicle and very functional. In engineering, the simple things are the best....

If "we round or soften" the corners of the caravan, we also get values of Cd = 0.6 (third case).

Cd

0.9

0.7

0.6

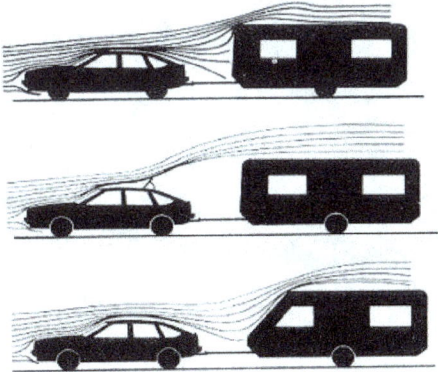

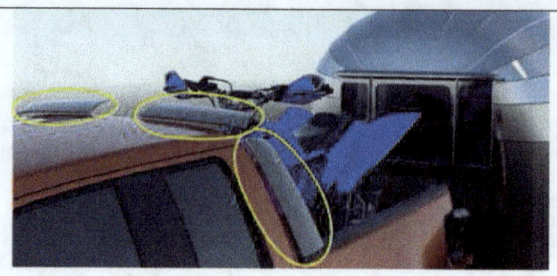

LOW CONSUMPTION AUTOCAR: XERUS OF TATA MOTORS

Tata Motors got in touch with us to design a low consumption bus; the goal was to design a whole bus introducing the latest ideas to reducing the aerodynamic drag. Later, when the crisis hit, also arrived cuts and the initial project so appealing was reduced to a re-styling of a bus already created by Tata, the Divo:

The project was also exciting, but was subject from the very beginning to the initial geometry of Divo: we could make it bigger, but not smaller. That is, we should base ourselves on its original structure:

We focused on four design premises:

- Bulbous bow or make the front section higher than the rear.
- Round front or nose.
- Adding a kind of wing-deflector at the rear (filling the rear zone).
- Entering aft ears.

Under these assumptions, we get drag reductions around 7% on Divo, but if not having reached the crisis, we had come easily to reductions up to 30%. Anyone is encouraged?

The next image, show the same idea about the "rear wing" or deflector:

The following image corresponds to the same ideas applied to an "ideal" truck:

INCREASE IN THE USEFUL POWER IN FORMULA DALLARA 3-306

Another example of drag reduction for greater top speed, we did with the Dallara Formula 3 car 306. Viewed from the front, we see that the mirror practically coincides on the side engine intake:

It is therefore logical to think that the rearview affects directly or indirectly on the engine power output. To know it we conducted a CFD study; simplification to be made, because at that time were not available full car model in CAD, admission, mirror and their respective positions were passed to CAD. Analyzing the pressure on the inlet, being and not being retracted the mirror, we obtained the following results:

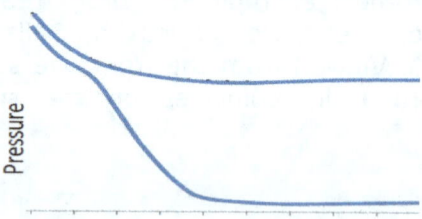

Clearly, with the retracted mirror, the pressure is higher; This is roughly equivalent to about 3 horsepower.

LENTICULAR WHEELS

In the Grand Prix of Turkey F1 2006, it was observed that Ferrari put a device on the rims of the rear wheels. According to FIA regulations, it is strictly prohibited the use of all types of attachments to the wheels and even more that they have aerodynamic influence of any kind.

Article 3.15 of "FIA Formula 1 Technical Regulations 2006, state that all parts of the car must be rigidly secured to the entirely sprung part of the car" y "must remain immobile in relation to the sprung part of the car".

Ferrari says it has only put this system to improve brake cooling, extracting warm air from them more efficiently.

The controversy lies right there: if this device that Ferrari used has also some kind of aerodynamic influence. In this small report, we shall see how this affects the aerodynamic device and to quantify the possible loss or gain is achieved, in terms of efficiency or resistance.

Many teams even that season installed similar things:

The basic objective of a wheel is the adhesion with asphalt, so that the maximum power transmission to be achieved by the motor. On this basis, its development involves various problems to be mitigated, but it is impossible to completely remove them. See each of them separately.

Friction with the asphalt:

It is a function of the characteristics of the tire and asphalt. By relying on the contact area with asphalt, is directly dependent on the properties of context, such as air temperature and asphalt, moisture, water or any other fluid on the track, type of compound, setup and some other element less important. This resistance will always exist and can be attacked only modifying and studying the compounds of the tires and car setup. Very often, increase this friction is not only convenient if not necessary. Generate more grip allows for example to increase the cornering speed and thus reduce the passing time.... In fact, friction or adhesion with the asphalt is the key in racing. The controversial Mass-Damper also acted on the same line trying to maintain constant grip during transitional phases.

It is according to the roughness of the tire rubber and the characteristics of the air; all these properties, in turn, as we have seen, depend on temperature, humidity, etc. In tests in wind tunnels before a race, you try to simulate all conditions of the race. Even simulating the roughness of the asphalt where it will compete. On the other hand, it is noted that because of the rotation of the wheel there is another phenomenon which intervenes directly on the frictional resistance which is the Magnus effect. This phenomenon implies a change in the boundary layer of air, directly affecting friction.
This resistance will always exist in greater or lesser extent and can be attacked only studying the roughness of the tires and trying that there isn't "too much" airflow in contact with the wheels.

Shape:

It is a resistance due to the very form of the wheel; It is directly proportional to the square of the speed, air density and the area; the proportionality factor, gives a coefficient of drag coefficient called "Cd"
This coefficient of resistance depends on the shape of the wheel, but also on the characteristics of the material. Logically, therefore, not much can be done about this force, if not divert the flow on wheels "properly".

Turbulences / low pressure or depression:

It is the resistance caused by the formation of vortices, eddies and turbulence in general. In the case of non lenticular wheels, the cavity of the rims and the passage of air from the inside, cause turbulence, which in turn, cause resistance. On the other hand, also all wheel produces turbulence in its wake; the larger these, the wheel will produce greater resistance.

Refrigeration:

Here is the key to the whole Ferrari device. In any F1 car front refrigeration intakes are very visible and air that enters them is "clean":

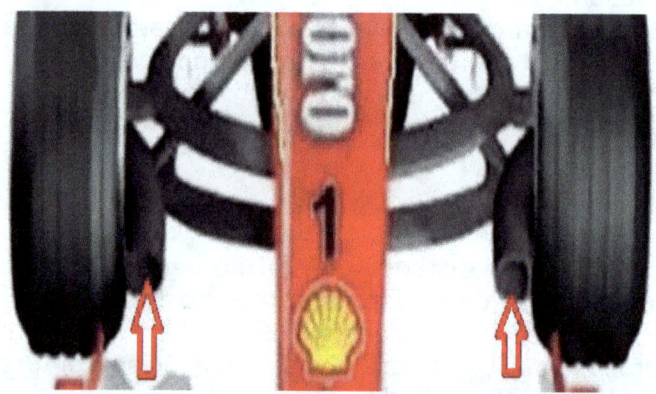

Not the case with the rear intakes that are "hidden" and whose inflow is "dirty" and turbulent; hence the cooling of the rear brakes is a more complex issue.
First, see if the placement of this device by Ferrari introduces some aerodynamic advantage, obviating the brake cooling. To make the appropriate aerodynamic test we will use two wheels.

First we will study over the rear wheel of a F3, later to make the study on the same wheel but making lenticular on both sides. The study will consist of a CFD simulation with the following information:

- Wheel speed: 250 km/h.
- Air temperature: 25º C.
- Real time calculation: 2 seconds (It is little, but the trend will continue).

Numerical data of both wheels are presented in the following table:

Lenticulares (Final step 0.019 seg.)						No lenticulares (Final step 0.019 seg.)					
Pressure Forces	PFx 161.36	PFy -1.268	PFz -1.3185			PFx 183	PFy 18.736	PFz -1.340			
Pressure Moments	PMx -7.794	PMy 131.31	PMz -1072.9			PMx -21.2	PMy 147.7	PMz -789.5			
Total Forces	162.21	-1.2649	-0.59723			183.75	18.738	-0.68609			
Total Moments	-3.0484	126.32	-1078.5			-17.857	143.16	-793.41			

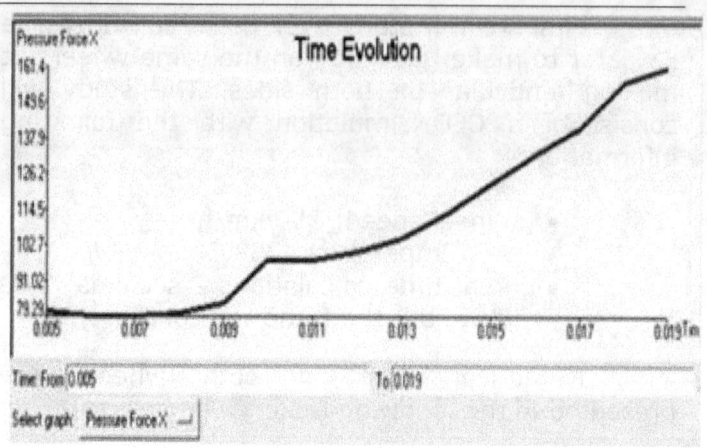

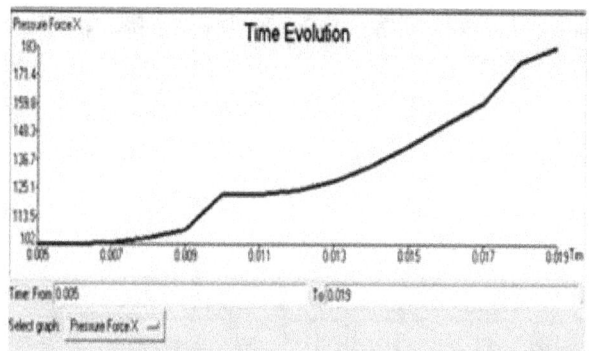

Looking at the data, we see that lenticular wheels offer about 8% less drag than wheels without lenticular (162 vs. 183), which is quite appreciable. We are talking in terms of total car the resistance is reduced by approximately 3%, which is also significant.

Anyway, the question is not so easy: from the beginning, Ferrari argued installation saying the device was generating more brake cooling:

See 2 different models of action in terms of cooling:

- Model A:

Through the "inside" intake of the wheel air is introduced at a certain speed and pressure. According to Ferrari device we can appreciate some kind of fan blades; it could be interpreted as an axial fan:

Therefore they induce us to think that it "helps" the warm air that has passed through the brakes to go out and dissipate heat:

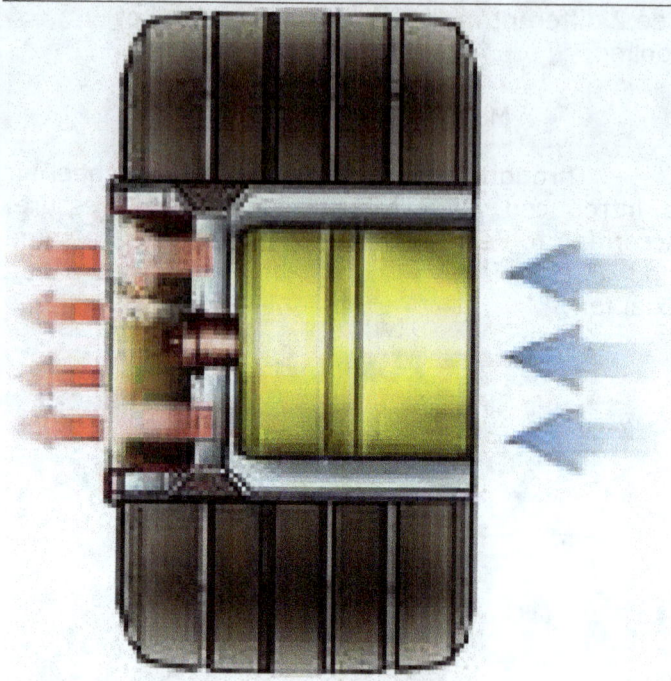

The "natural" tendency of the flow is to go from the inside out, because on the outside, there's less pressure than on the inside and also the air enters the cooling intakes from the inside.

The higher speed, higher flow rate has to be removed:

CAUDAL

VELOCIDAD

Anyway, this helps to the extraction and heat dissipation, also involves a loss of power (increased friction, flow deflection, etc); power depends on the cube of the speed, so said this, making the appropriate and necessary calculations (knowing the efficiency of the blades above and other data), we know the power loss due to aid removal.

It would be possible to find with the surprise that the power loss due to this "help" is greater than the drag reduction due to the fairing of the outside of the wheel .... But we also might find that Ferrari wins top speed, since it is possible to win in efficiency by altering or varying the cooling flow and the gain or improvement is greater than the aerodynamic lose.

- Model B:

Looking the "blades" of the device (or at least what seem blades ....), And the theory of mechanics of "traditional" fluid, it is a centrifugal fan; thus, the air does not exit through the center hole that is observed on the rim, but it would be entering. This would be consistent with the fact that Ferrari:

- It has removed internal cooling intakes or has been replaced by another "system", perhaps centrifugal.
- They haven't removed them.

If they haven't removed them, look at the following: if the inside takes exist, the air is introduced into it; at high speeds, the centrifugal fan will introduce large amount of air, so that it is possible that the balance is positive outward or inward, but not widely; this would cause the brakes to cool less at high speeds (i.e. straight) and cornering; therefore would make this a very efficient system because the heat would keep the discs to be exploited "conveniently"; one of the problems that always exist in circuits with large straight, is just that: the discs are cooled too much to work well at the end of the straight, cornering or as needed.

Conclusions:

As seen, all are speculations and hypotheses of what could be and behave. We do not have the device in question in our hands, which something else is not possible. With all that said, Ferrari may lose aerodynamic drag and thus gain top speed, but also may even lose top speed; everything depends on the balance of power gained and lost. However, we have also targeted and outlined a new cooling system at least at first glance, it seems viable.

In conclusion we can say that in terms of cooling efficiency, Ferrari has successful with installing this device and somehow prefer this, to to increase the top speed; although it may even increase.

The possibilities, as we have seen, are many and varied. In short, if we look just at issues of fairing hollow wheels, the resistance is reduced significantly; but we must not forget that the brake cooling is essential.

In addition to brake cooling, there are several methods to channel air through wheels, to adapt their flow to a certain area or a certain goal:

The wheel aerodynamic influence is evident; consider the following study to see their influence on a touring car:

Suppose the Audi A3: Let us see a small quantitative study of the influence it has on the Audi incorporation of wheels and a cover-step wheels; We conducted a series of studies based on the own model Audi unmodified; see the "CD" in terms to put or not put wheels and careen or not careen wheels:

	Cd	Cd	
CON RUEDAS	0,315	0,315	CON RUEDAS
SIN RUEDAS FRONTALES	0,274	0,303	CARENA RUEDAS FRONTALES
SIN RUEDA TRASERAS	0,26	0,31	CARENA RUEDAS TRASERAS
SIN AMBAS RUEDAS	0,215	0,295	CARENA AMBAS RUEDAS

SHARK FIN

Conceptually, it comprises a flat surface that properly managed the flow into the wing and divides the flow in two halves.
This is the first application. However, taking into account this objective, it works in two ways:

On the one hand:

- Divide the car in half with what "physically prevents" the flow to pass from one side to another, mitigating the turbulence that occur behind the intake.

- The sliding flow on the fin surface and due to the "Coanda effect", is adapted to the sticking surface. It will increase the frictional drag but surely the benefits achieved in terms of lift will be higher, to improve performance and area of action of the rear wing.
- Given the location that have the fins may be able to reduce the total resistance to mitigate the turbulence generated behind admission.
- In corners fasts and mediums, produce undesteering.
- If the rear wing is lower, the shark fin is more necessary.

Thinking about other possible applications or functions that this flap may have, we can think of:

- Stabilize the dynamics of the car in curves.
- Place the global and lateral center of pressure of the car where required.

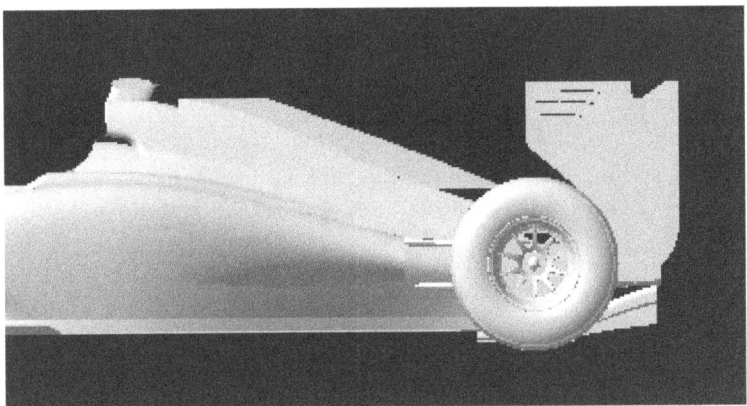

The main reason is the rules changed the position of the rear wing, making it lower. In this position the wing is no longer jacked up high above the car, the airflow reaching it is now obstructed by the roll hoop and the general airflow passing closely over the car's bodywork. With this dirty air the wing is less effective and thus it needs to be set at a steeper angle of attack to make the same downforce level. With this wing angle come drag, along with the huge new tyres, this will really slow the car along the straights. So, the teams want to smooth the airflow towards the rear wing. As the rules restrict where you can put bodywork you cannot have a horizontal vane\wing ahead of the main rear wing to clean up the airflow like they had in 2008. So, the only option is to use a shark fin. Also is possible to use it for refrigeration help:

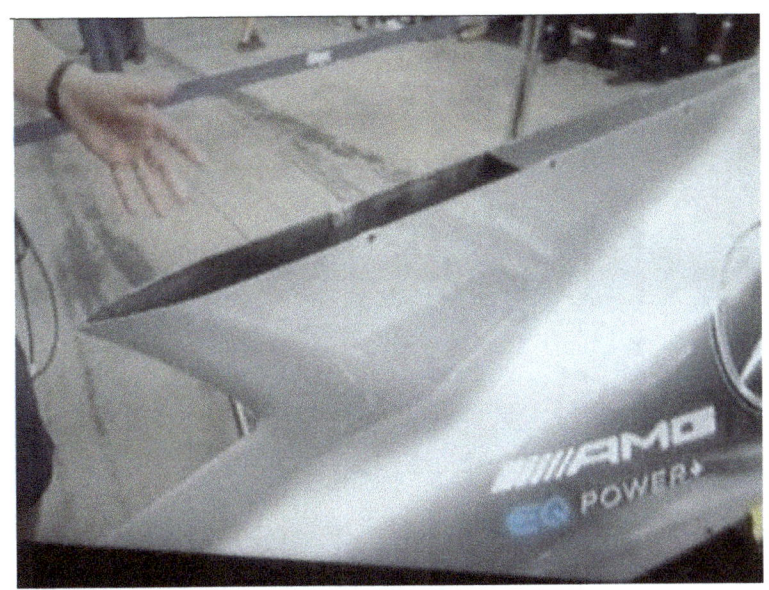

In action:

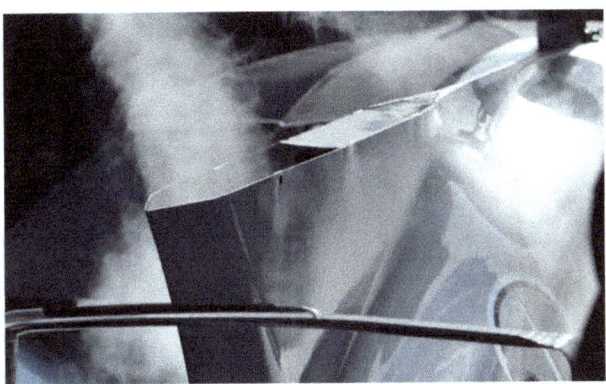

MCLAREN FRONT WING (WING IN BOW)

Technical disquisitions:

Whenever you ask what it is for determined aerodynamic device, we can only make assumptions, since the exact reasons can only know engineer itself or the tean who designed the item.

These possible reasons and functions of the wing could be:

- Deflection.
- Increasing Downforce.
- Optimal flow channeling aft.
- Increasing engine power.
- Wing for acceleration.

As will be seen, some may become counterproductive.

Technical Regulations are clear: any element must flex a certain amount or less.
For what would be wanted to flex this device?

In the case of the lower plane of the front wing, which is closer to the ground, it is clear since the closer to the asphalt the more downforce the car generates asphalt.

In the case that concerns us, the question is more difficult to explain because if deflected from its central part, the downforce does not directly increase (although useful in acceleration for example, when the "other" lower wing does not function properly because they are very separated from the asphalt). You could optimize the deflection causing the leading edge down and the trailing edge rise and thus increase downforce, but this would make more dirty air goes to the back.

Note also that other teams have repeatedly adopted similar solutions to this; the question they all had a center support to prevent excessive deflection and therefore punishable. By not bring central support the Mclaren system, it is conceivable other things.

The fact that bend at the central part without changing the angle of incidence, only modifies the flow aft, not producing itself increased downforce.

The fact that the wing vibrates can be beneficial to grip with asphalt, it can act as a kind of Mass Damper; must effectively control such vibration and "dock" to the vibration produced by the suspension and other parts of the car.

If we add to the above the observation that the wing has no central support, we will realize that what is being pretending is to interfere as little as possible to the flow into the intake port; the fact this support exist adversely alter this flow.

Conclusions:

Therefore and as a summary, I would say I lean more to think that this is a streamlined solution which aims to increase engine power, adjusting the flow into the intake port. It was leaked to the press directly Mclaren that this wing increases to 7 hp engine power.

REAR WING SUPPORT LE MANS

In virtually all formulas or categories where rear wings are permitted, the method of anchoring them to the chassis is simple: through vertical columns that connect the wing to the chassis:

In most are choosing, if regulations permit, to unite the two parts by a curved brackets:

These supports are less affected to the function performed by the wing reducing the drag that originates and increasing the efficiency of the wing; these are the two advantages of using this type of support.

We can see this pressure variation, from the next CFD image (here is interaction between wing and support):

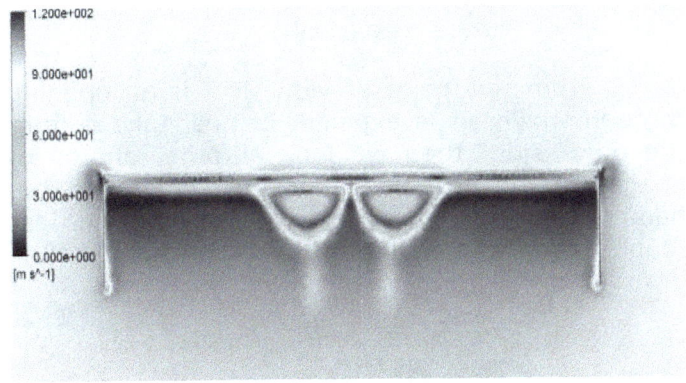

FERRARI'S TUNNEL NOSE

The first descriptive picture that we saw of the so-called new Ferrari's tunnel nose was as follows:

It is clearly appreciated and according to the position of the front wing, a plurality of apertures. Considering this photo, could be of 2 types:

- Chimney type (extraction due to venturi effect).
- Nozzle type (push).

From this point of view, if it is an opening through which air is expeled, it must take it from the other side; then we look at other photos or images of the new nose: we get something that had not been seen and it was neither more nor less than to know where to it took air:

It is seen perfectly as indicated: there is a lower opening through which air is drawn. We have defined the application suggesting this design.

On the one hand:

- It produces downforce, since it diverts certain airflow up:

This is the first application or effect of having such design; The following result is that more downforce is produced, so is not necessary that the front spoiler is so large so that there will be less resistance and hence the top speed will be higher.

- Moreover, large openings which are

located on the top of the nose are due to extract as large airflow is not an easy task; It is to channel the flow of an ideal way; so does diffusers fins for ground effect works best.

It is a self-generating system; that is, a system that acquires the optimal operation conditions for it; due to the arrangement of the openings on the nose, a depression or low pressure area which sucks air is located in the bottom opening of the nose occurs; the more the car speed is, the greater this pressure difference and therefore the flow of circulating air.

We know that there is a fundamental and basic principle of competition, which is to use to "something else" all existing elements on a car, in addition to the function for which was laid or created. For this reason, we can think of three more functions, due to the position they have outlets on top:

- Conveniently adapt the flow through the bottom of the car; We imagine the effect that this floor is optimized and therefore the downforce generated: the lower intake reduces the amount of air passing under the car.
- Increase engine power.
- It reduces the resistance of the nose, as it eliminates the boundary layer that forms in front of the air intake, and prevents a boundary layer is formed too big behind it.

This latter aspect was already discussed by Mclaren on his new double front wing. Ferrari may do the same with this tunnel.

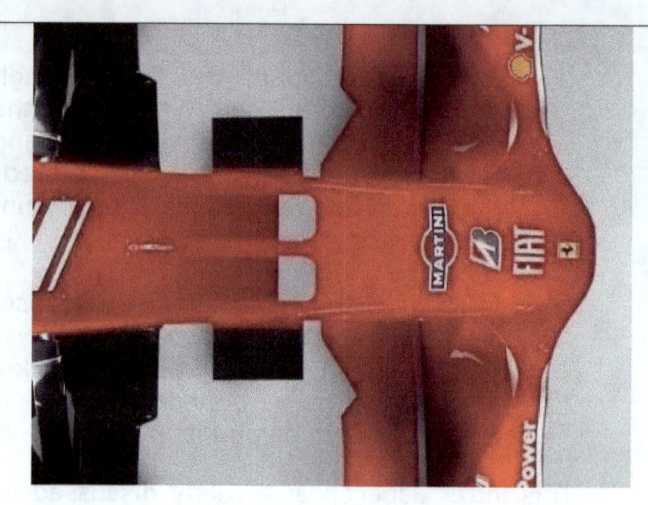

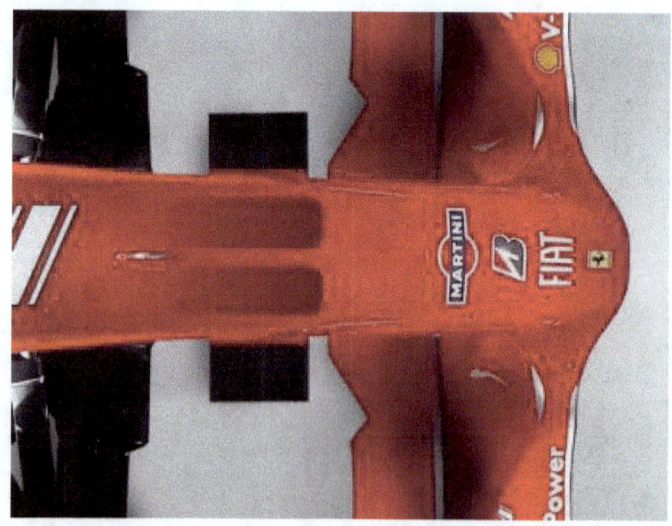

This system, exist a lot years ago (Ferrari, Ford, etc....):

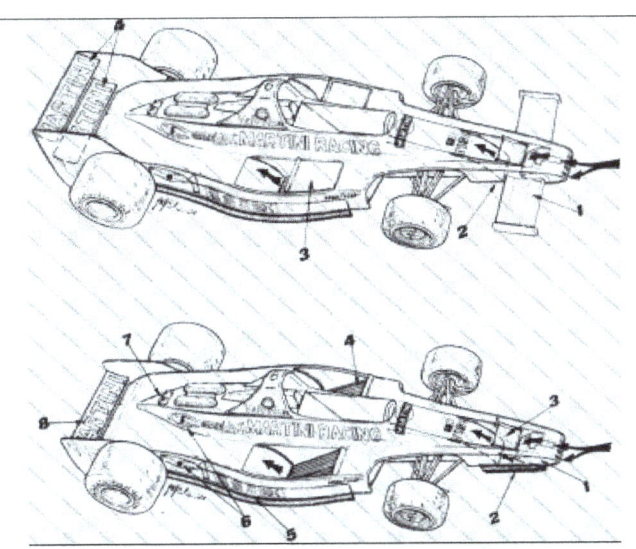

AIRBOX AUDI 1600 RACE CAR

When we considered this study, we had to start by finding out if the air distribution to 4 piston was optimal; We received the design directly from Audi:

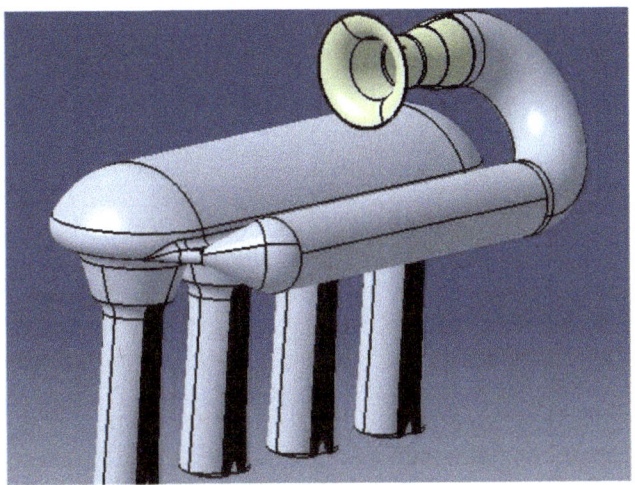

It was a flange or inlet restrictor, followed by the airbox to the engine itself.
We chose the following method of study to reduce calculation time:
Test the inlet flange calculating the maximum flow coming out from the flange; thus we took this data as initial input flow to the airbox to do a posteriori the other study of dispersion in the airbox.

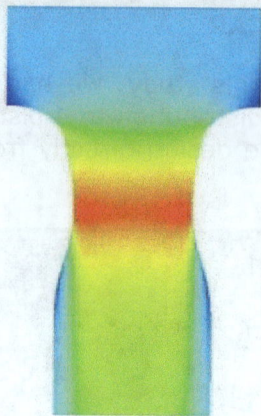

Maximum mass flow rate: 0.385 Kg / Second.

Making a cross section on the airbox the pressure distribution, velocities and flow inside the airbox can be seen; but what we wanted was to know the flow rate each cylinder could collect; the results were clear: there were two cylinders and especially one which received little air compared to the rest.

Conclusion: bad original design; it was necessary to stop and redesign the airbox.

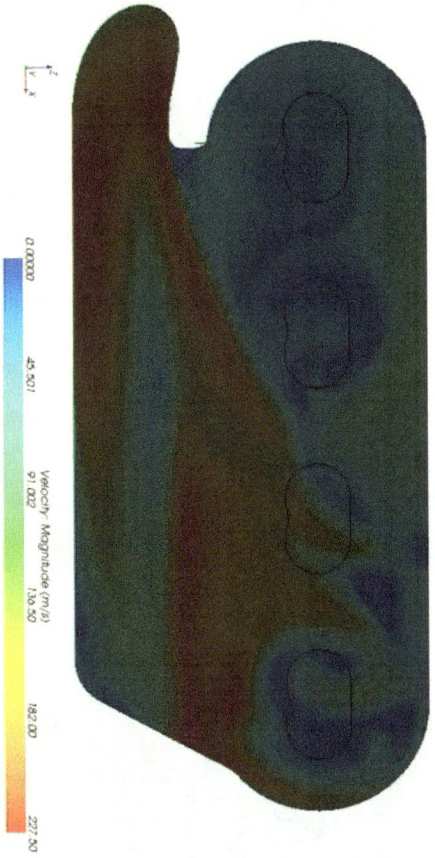

Velocity Magnitude (m/s)

0.00000   45.501   91.002   136.50   182.00   227.50

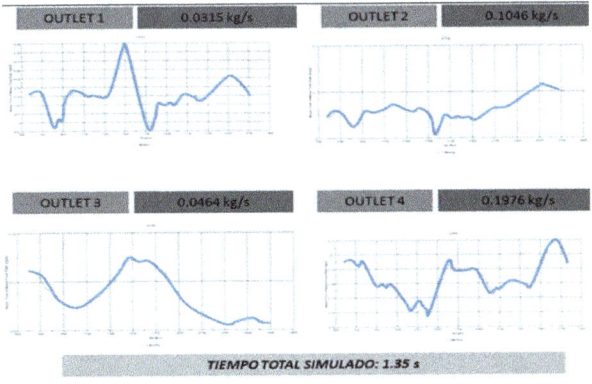

| OUTLET 1 | 0.0315 kg/s | OUTLET 2 | 0.1046 kg/s |
| OUTLET 3 | 0.0464 kg/s | OUTLET 4 | 0.1976 kg/s |

TIEMPO TOTAL SIMULADO: 1.35 s

APRILIA 125 CC AIR INTAKE

Aspar Team had doubts in 2007 on the efficiency of its 125 cc bike since according to them, the bike was under the bike top speed of 125 cc, 2004.

The only noticeable difference was the disposition of the airbox; in 2004 it stood frontally, while in 2007 it was laterally:

At first we thought that the "failure" was in the airbox itself; we tested both systems by CFD. The results corroborated the pressures in the carburetor were the same, indicating that the fault was not there:

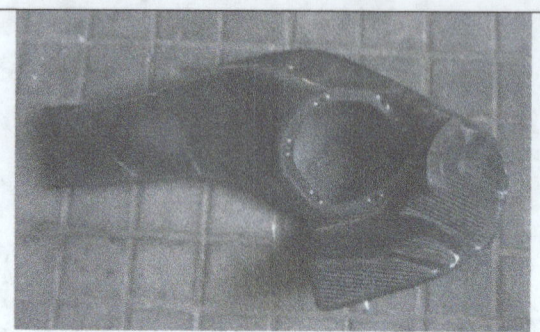

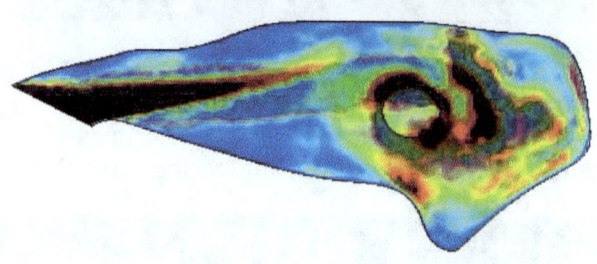

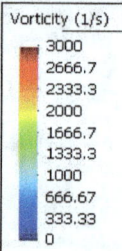

Vorticity (1/s)

3000
2666.7
2333.3
2000
1666.7
1333.3
1000
666.67
333.33
0

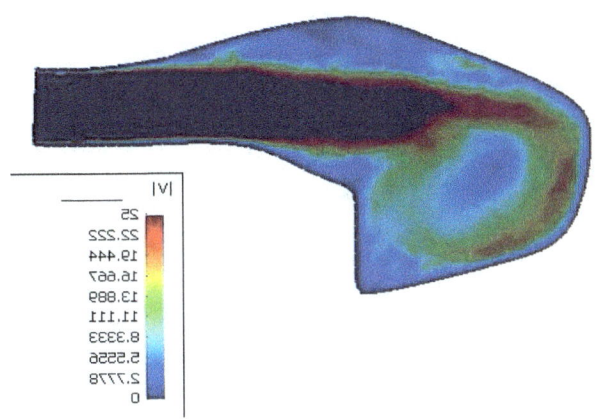

We passed to study the bike as a whole; CAD design was performed using photogrammetry and Rhinoceros; the results denoted that what failed was the position of the airbox inlet; It was more efficient the intake of 2004 than 2007.

The following year, Aprilia changed again the position of the airbox ....; it was also used a reduced or simplified CAD model of bike:

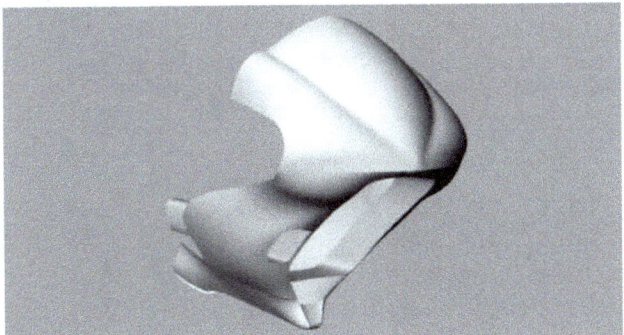

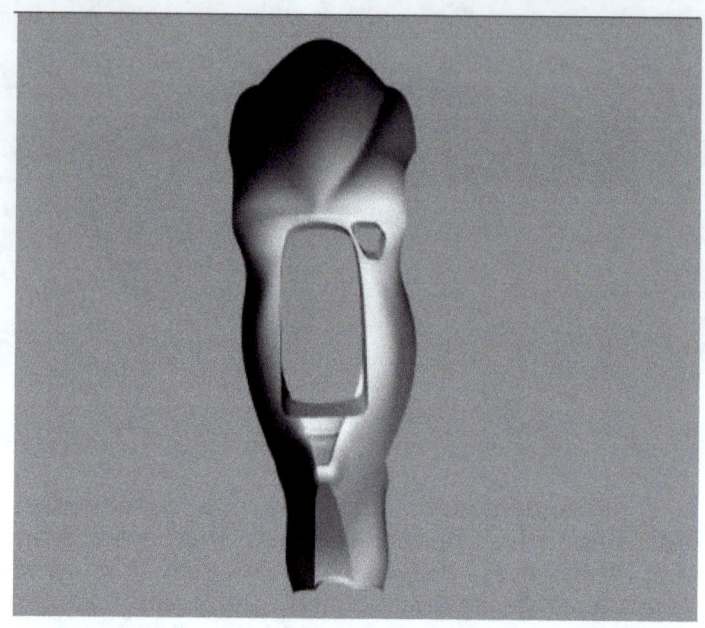

With the arrangement of the airbox as 2004, the bike was about 5 km / h top speed:

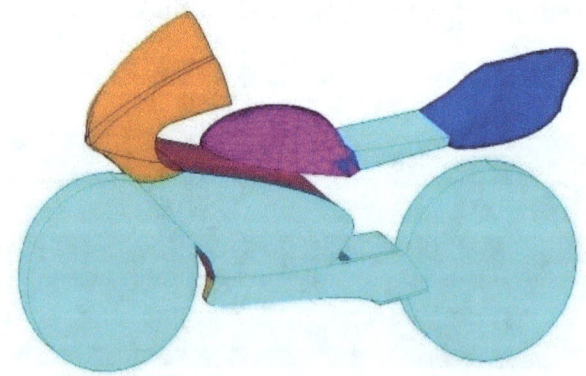

We also measured the pressures in both models of motorcycles; for it and as Aspar Team did not want any element of their motorbikes were used, it had to design a system including battery, sensors, data acquisition and processing:

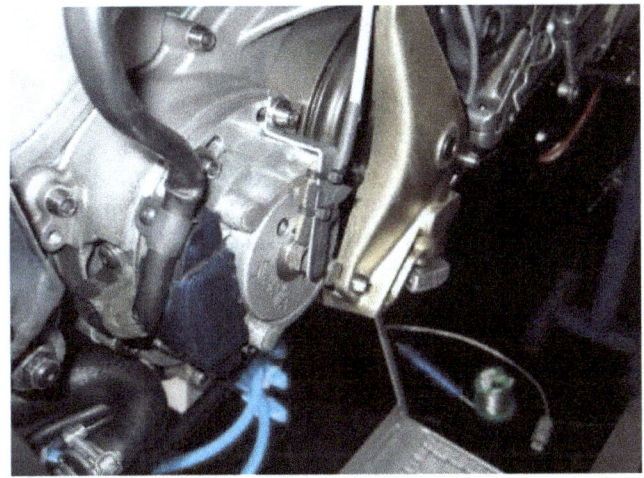

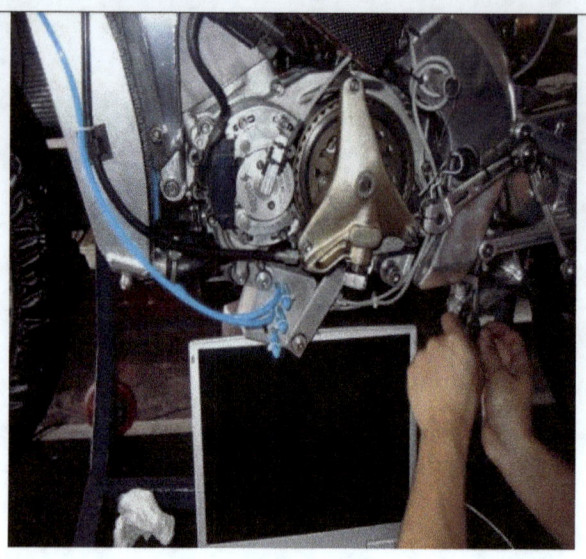

DALLARA 306 FORMULA 3

This was to analyze and calculate the center of pressure of 306 Dallara Formula 3 in 2 different configurations.

We needed to have the car in CAD format suitable for a CFD trial. And that for budget we could not generate and use a 3D scan and also because the technology was beginning, we chose to use the software Rhinoceros; we had many compatibility issues in geometry as IGES generated by Rhinoceros were not to the liking of Gambit (mesher of Ansys).

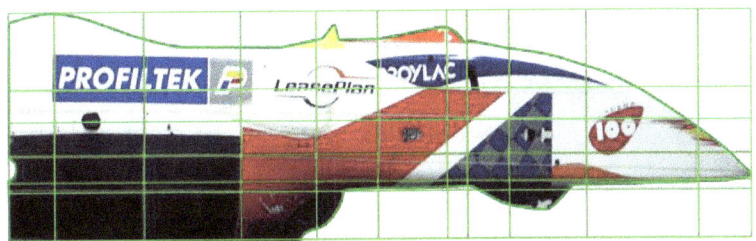

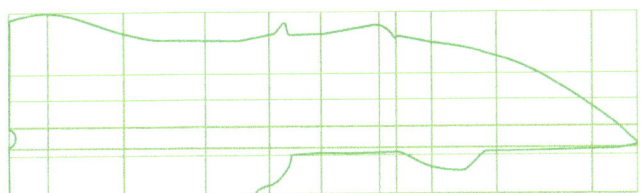

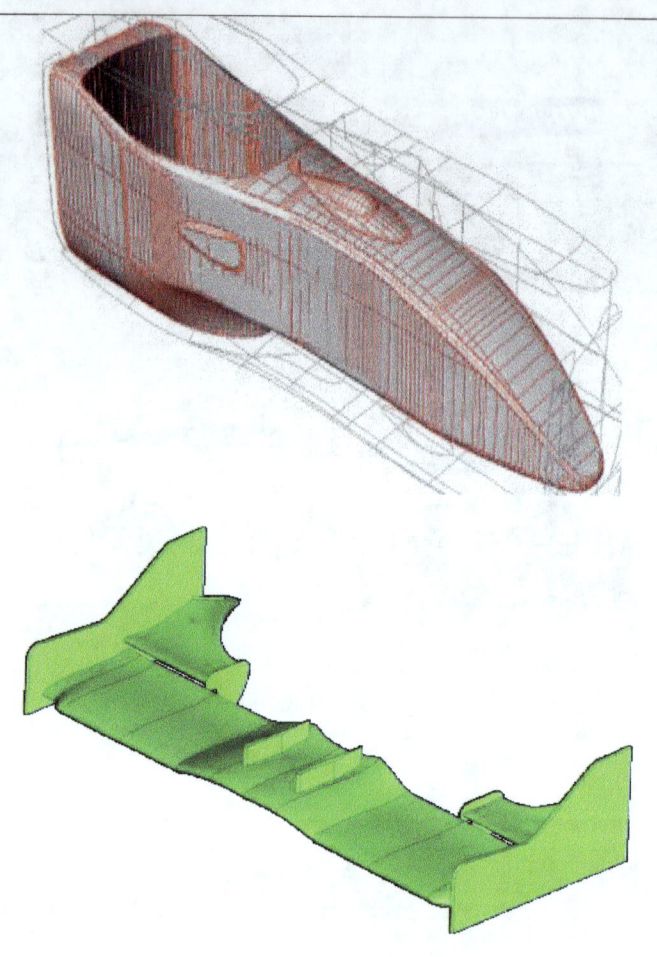

By the method called photogrammetry and piece by piece, the car took shape; photos from different angles and from a distance to avoid the problems of perspective were taken for it; from these pictures, it was generated the CAD of the car:

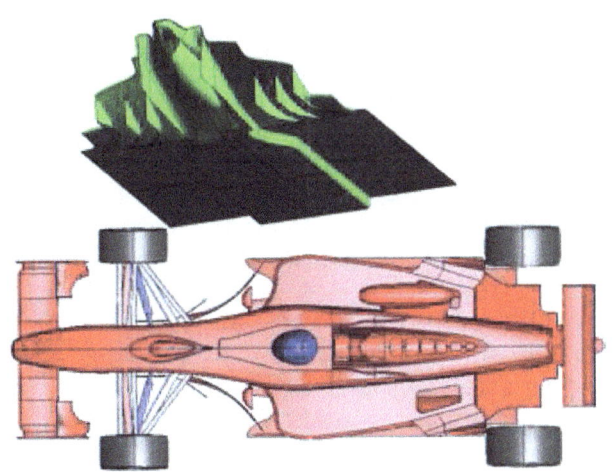

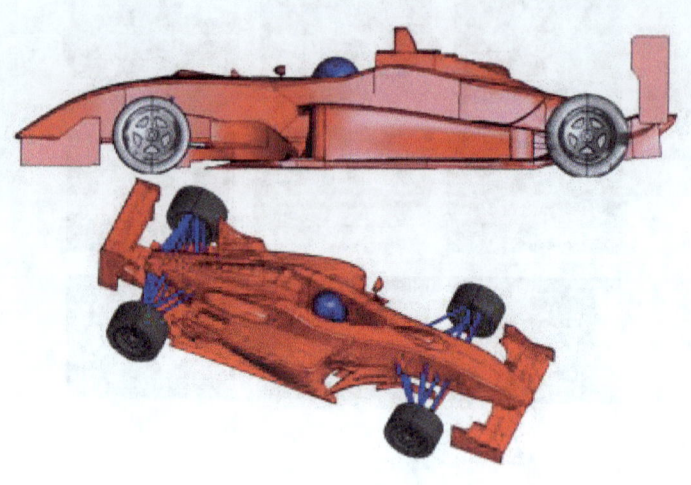

ESP: ELECTROSTATIC PRECIPITATOR

This time, it was to optimize the performance of a device, called so; It is a system that is responsible for cleaning the smoke of solid particles. The work was done for the company Johnson Mattey.

The system, from a vertical plates electrostatically charged, causes the dust from sticking to said walls and by vibrations, falls into a series of hoppers located at the bottom.

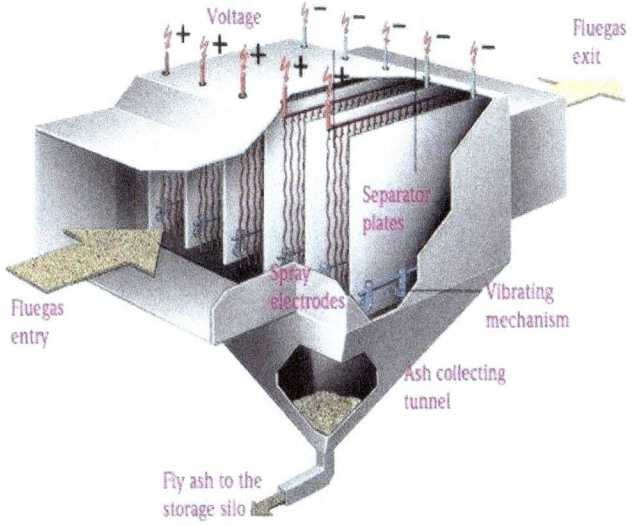

Voltage

Fluegas exit

Separator plates

Spray electrodes

Vibrating mechanism

Fluegas entry

Ash collecting tunnel

Fly ash to the storage silo

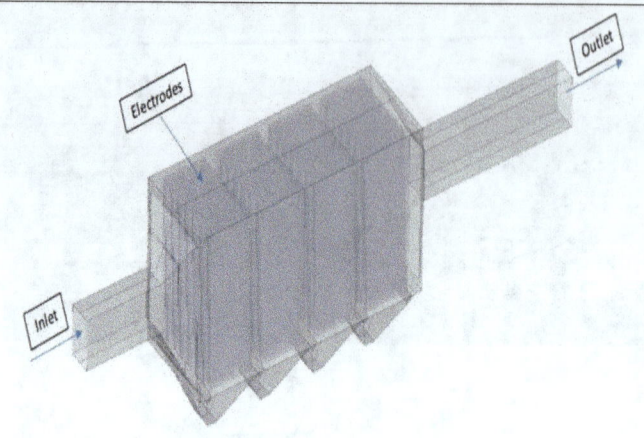

The aerodynamic problem lies in the fact that the inlet duct is small compared to the section of the apparatus, thereby electrostatically charged plates are not exploited to its full extent; to view this and make a state of the art, we studied that "already created".

The diffuser standing at the exit to the chamber where they were the plates was:

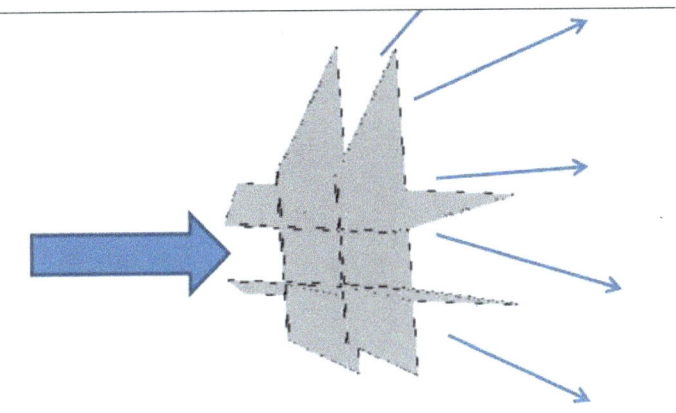

Note that if Bernoulli be fulfilled in reality, we would not have any problem because the flow is evenly distributed throughout the section, however great were the differences in diameters.

But this was not the case: the system does not took advantage of the full extent of the charged plates:

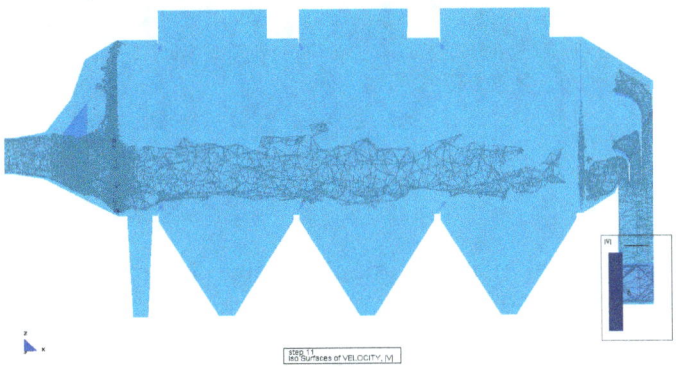

To improve the system, we came up with countless possibilities:

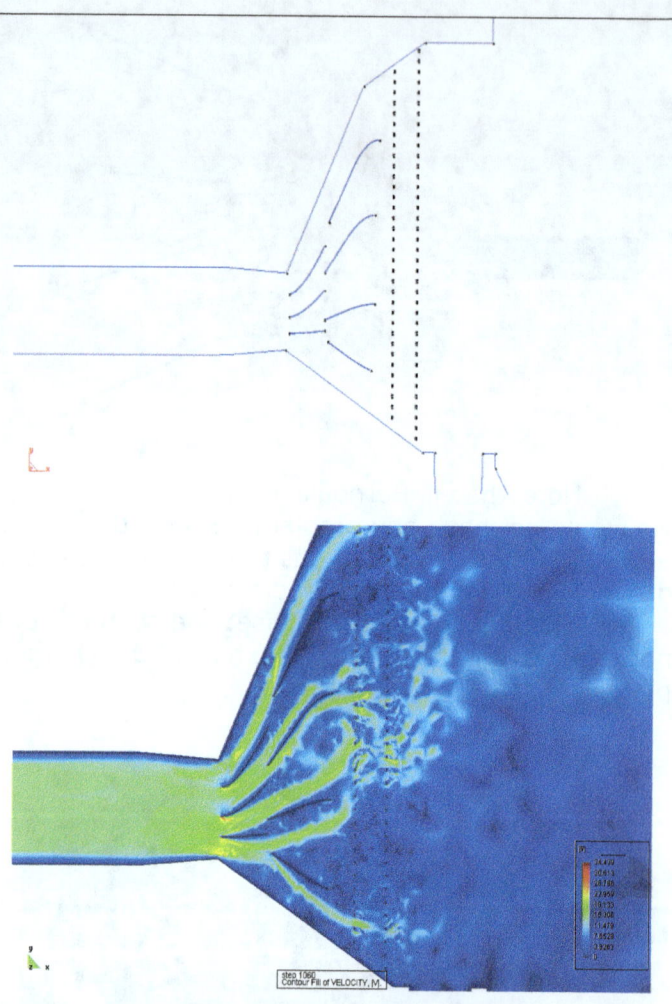

But neither significantly improved the system or the original deflector.

If you look at a speaker, we see that it has smaller holes in the middle and are larger as we move away from the center; and the sound is distributed in all directions; this was the essence of our new idea deflector:

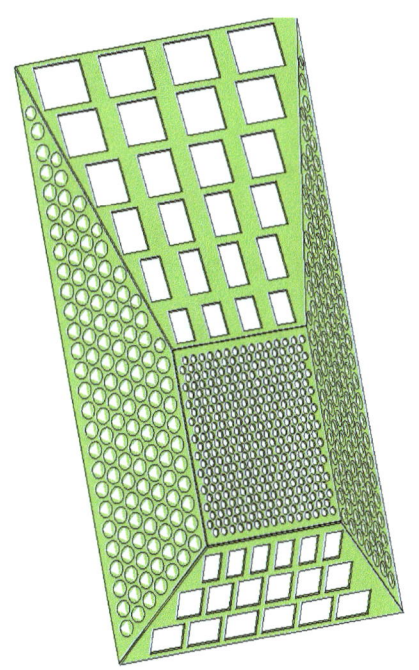

And the system, as shown, works very well, dividing the flow across the chamber:

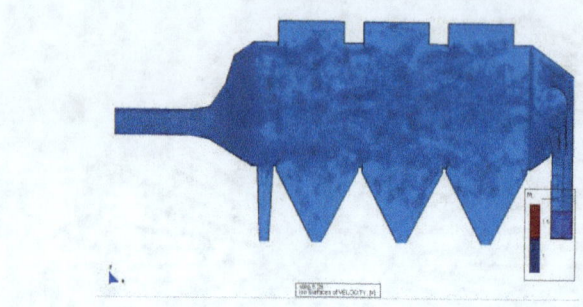

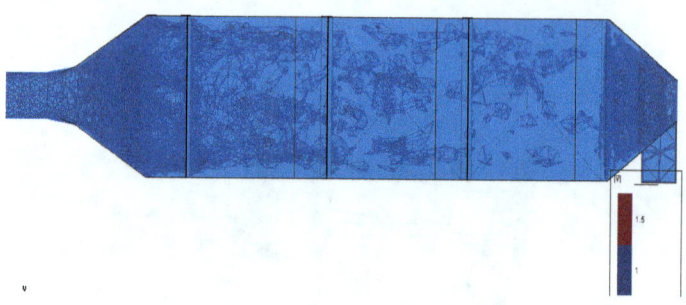

To round off we carry out "real" small scale studies (transparent walls), in order to not only refine the numerical models used in CFD simulation, but also to be able to, depending on the dynamics of the air, think of other possible solutions and even test them during the tests:

MOSLER MT-900-R WHISKERS

This study was based on the experience and trial and error tests, typical in engineering; this choice was primarily due to three factors:

- We did not know the existence of 3D Scan or anything.
- We did not have helpful budget.
- The urgency: it was a Monday and the race was on Saturday.

The process was simple and effective, but not "ideal" or fully optimized: from ideas we proceeded to manufacture them and install them on the car and we chose the best option that was the one that produced more front downforce; tests were performed on track, with elements made of riveted aluminum plate. They were made on Thursday and Friday before the race...

DRAG REDUCTION SYSTEM IN TRUCKING TRANSPORTATION

The aim was to design a system placed in the upper back of a truck to put air in the depression area of the stern; To do this, we use CFD techniques on a truck as simple as possible; simulations clearly displayed the objective. It was not at all optimized for urgent manufacturer-customer (SDR system), thereby reducing fuel consumption was just 1.5%; making further changes, we have achieved close to 12% reductions; pity that these new versions are not marketed; anyone is encouraged?

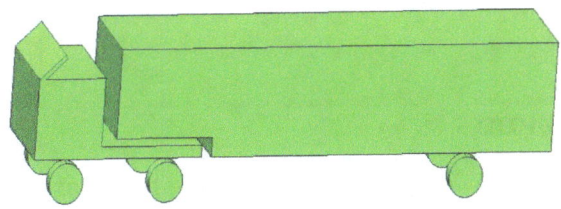

With and without the system; the air gets before in depression:

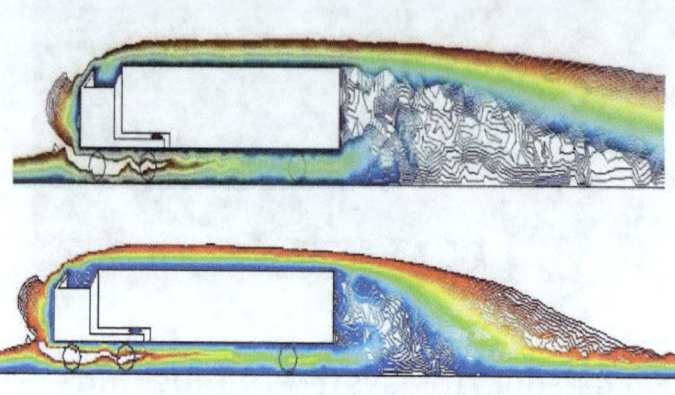

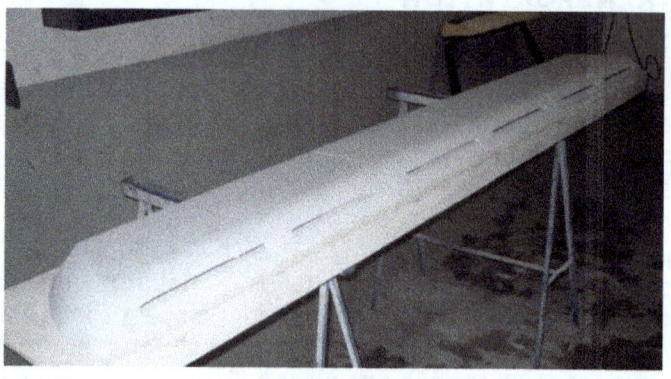

SIDEPODS

The study tried to find the best choice among:

- Place bargeboard.

- Design the best sidepod geometry.

The efficiency was based on criteria of downforce generated by the sidepod and the mass flow that could pass for it in terms of energy loss of the radiator.

On a "base" sidepod were made 3 models and mass flow, pressure and downforce were calculated; Based on these results, the best option was chosen:

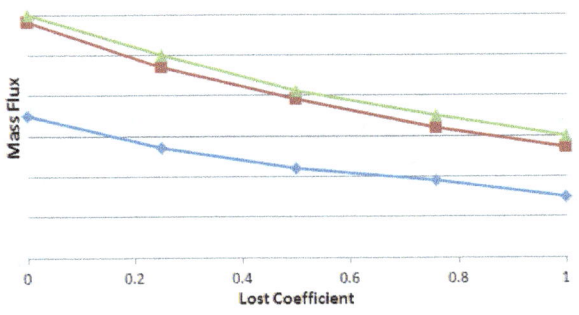

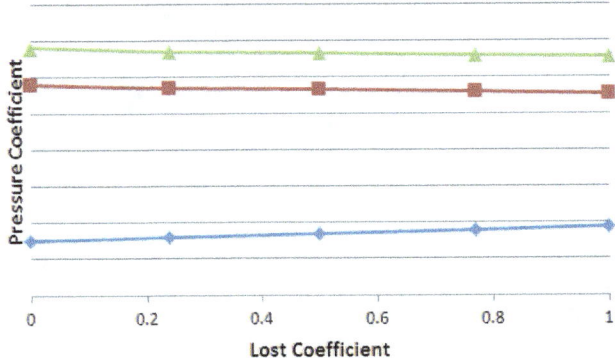

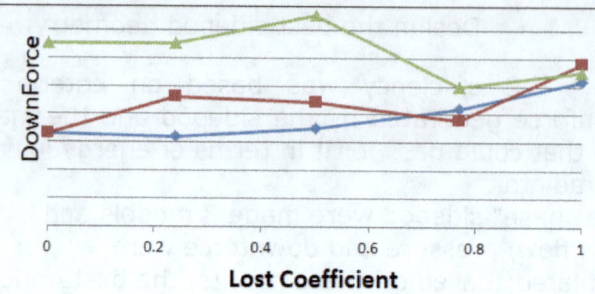

The sidepods must exist; the evolution that have suffered throughout history is evident in the following images:

PAINT OR ROUGH SURFACE; TURBULATORS

We already know that turbulent flow tends to adhere more and better on a surface that has a high angle of incidence, near the angle of stall.

In the case of a diffuser we could make turbulent flow was initiated just before the diffuser; To do this, we use rough painting to achieve this end. We do not intend to give rise to "cheating" but it is very important to skip the rules "legally"; this implies not only fully meet the regulations, but learn more about the different aerodynamic methods you can use.

You can use rough painting and even unions adhesives obliquity, as turbulators....

DRAG REDUCTION AND OPTIMIZATION OF A GURNEY FLAP: CASE OF DRAGONFLY

We know and we learn from nature and as stated earlier, in particular animal.

On this occasion we analyze which system might be the best when it comes to design an effective Gurney flap.

We started doing various studies several gurney flaps, obtaining that the optimum is:

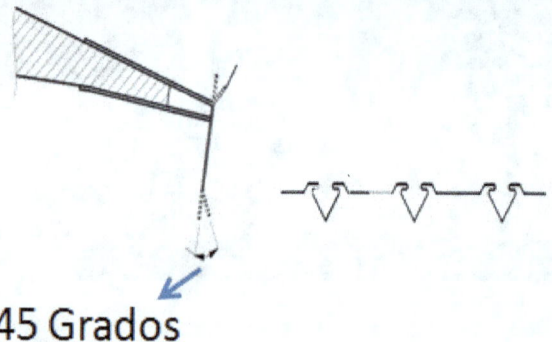

45 Grados

Look at Dragonfly:

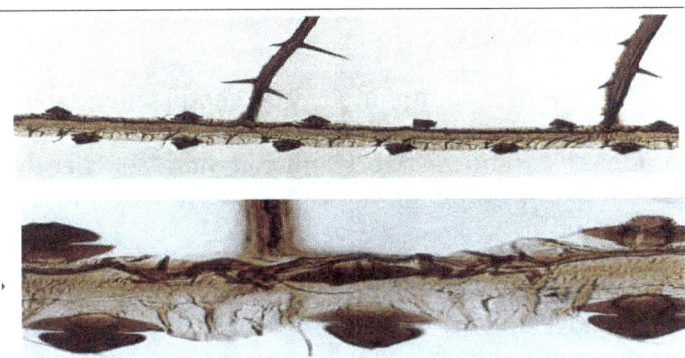

True that if we worked harder the animal world we would have been able optimize "quickly" our gurney ???.

The similarity between "reality" and previously achieved is incredible; must take into account something very important:

The area corresponding to the boundary layer that we know is only a few millimeters above the surface area is the area where the most important movement in relation to the future dynamics air takes place; what happens inside, completely determines the overall dynamics of air and dynamics of the particular car;

Therefore, whatever we do within this layer is crucial to the overall dynamics.

Do not let small "gadgets" or "systems" that "seem" to alter "little" the flow aside, it will just look that way.

Do not think only in this example of dragonfly, but also in shark skin, etc....

BLOWN EXHAUST

We saw something in the section concerning methods of reducing drag; Brawn double diffuser made the diffuser work much more efficiently; blown exhausts also fulfill the same function as the double diffuser.

In the 2010 season exhausts were allowed close (just 30 cm) of diffuser; This allowed Red Bull to use the exhaust gases to increase the efficiency of the diffuser; this is performed electronically varying the engine map such that in slow corners, where speed is required to the diffuser work properly, the motor speed up (without driver causing it) to inject high velocity air right on the top of the diffuser.

FIA changed 2 things from next year:

- Advancement of the exhausts almost one meter.
- It banned the "electronic" engine acceleration.

These events forced Red Bull to manage to carry the flow exiting the exhausts to the rear, so that would work analogously as in the 2010 season.
It was based on the Coanda effect, in making holes and slots acting as suckers of flow, in the use of turbulators and Gurneys that created depressions, etc....
The vertical piece that is placed at the beginning of the sidepod, is the first to help the flow to adhere to it:

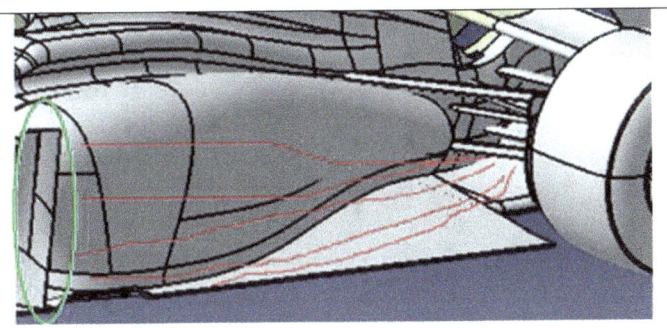

Red Bull has incorporated a "crack" or area of suction to suck the flow coming from the sidepods, helping to move toward the top of the diffuser:

It has also been adopted by practically all teams; pursues the same objective: to optimize the performance of the diffuser. Since the distance between the diffuser and the exhausts is so big now, teams must think efficient methods to communicate "exhausts - diffuser".

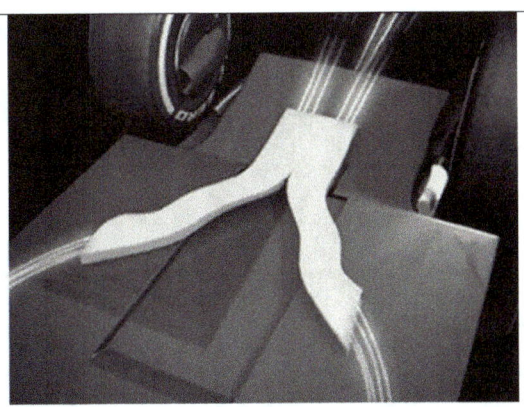

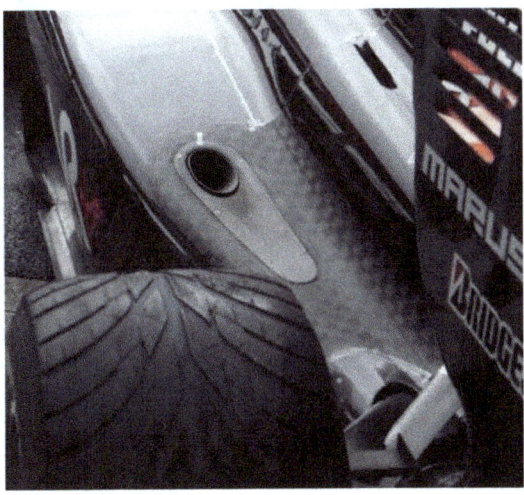

Another method for improving the difusser (helping the floor), is blowing bottom:

SLOTS ON THE FLOOR

In the 2012 season we see the incorporation by Red Bull of a series of openings in the floor; they were located in the rear, before the wheels. Its operation is threefold:

- Communicate the top with the bottom; through such connections air passing over is aspirated; this suction creates an airflow "pushing" the front air toward the inner rear. It is in short, to help the flow of "main" air to flow toward the rear and optimize the performance of the diffuser.

- Inject air just ahead of the rear wheels to reduce drag taking advantage of low pressure.
- The airflow is introduced into that area, it is deflected by the shape of the orifices or cracks themselves, towards the rear of the diffuser, in order to increase the amount of high velocity air and increase the efficiency of the diffuser.
- Help to diffuser work better; there is then, a flow air to diffuser.
- To prevent the "Aqua planning"….

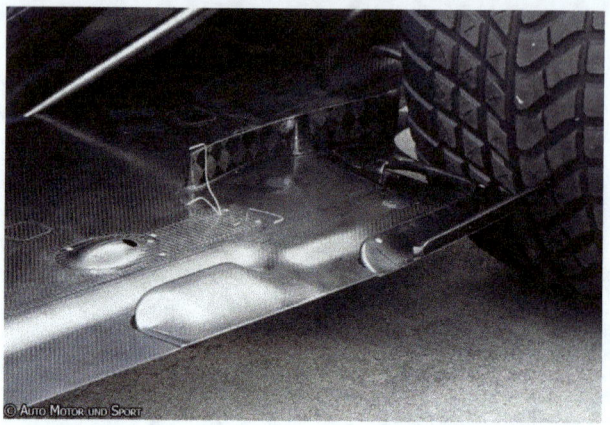

This method of communicating low with high pressure is very useful and can be applied to many cases:

- Help the Coanda effect, as in this case.
- To reduce high pressure, reducing drag and thus may increase the top speed.
- Help other effects or cuts to do what they have to do and improve their performance.

The fact relieve pressure in general has many benefits that we have seen repeatedly throughout this book.

Here we see a glimmer of the Regulations and the teams have taken advantage for these openings; the issue is that the legislation says that in a certain millimeters wide floor can not practice holes; It is that this width is smaller than the width of the floor; left over 5cm on each side; it's just in those bands of 5 cm where openings, gills, holes, etc are concentrated.

TURBULATORS ON SIDEPODS AND CHANNELS (MCLAREN 2012)

In the Grand Prix of Germany 2012, McLaren introduced a series of flaps at the top the sidepods; anyone would think that are responsible for directing the air towards the rear in an appropriate way; but this really works well in combination with the Coanda effect, the flow that goes above the sidepods must traverse a "long distance" to reach its destination; is better and more convenient converted into turbulent flow; the reason already know: tends to separate less from the surface of the sidepod and thus more easily channeled:

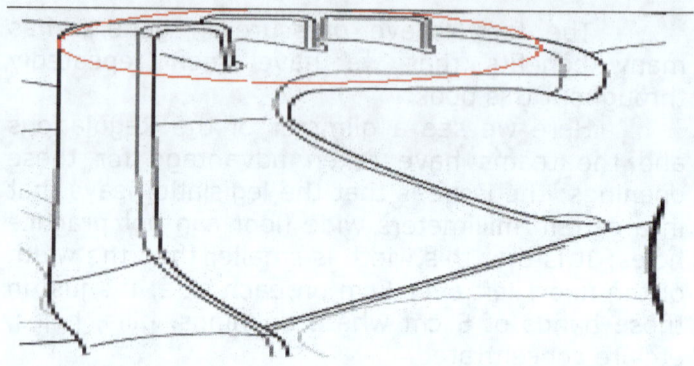

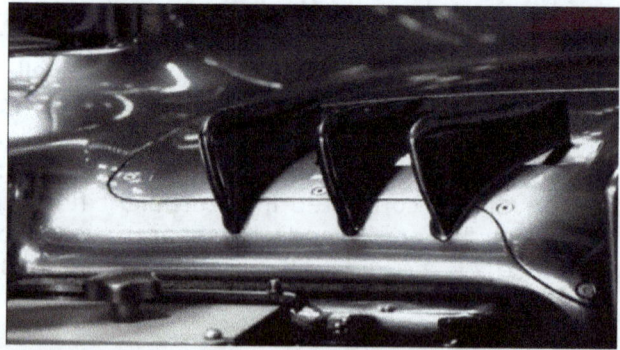

Another important reason lies in the fact of deflecting the airflow of the rear wheels; it reduces the overall drag of the car very appreciably.

In the next picture, do not you notice a striking resemblance between the turbulators of the airplane wing with Mclaren turbulators? -- Remember please, as we know already, that these turbulators are also vortex generator in order to help air to go where is needed (turning sense determined, to up diffuser).

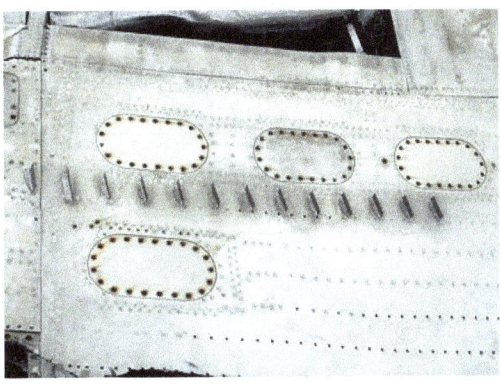

In the race of Spa 2012, Mclaren chose another technical solution but just the same background and objective: make the flow to "stick" to the surface of the sidepod to reach as far as possible and channel it toward the rear and / or diffuser; this system can be a great help to the exhaust, to go where they go: the rear:

There are many variations of the same system:

The airflow is accelerated by reducing the section making the air to have more energy, longer adhering to the surface; on the other hand also they play a role in helping to Coanda effect: to be located where they are, become the first aerodynamic piece that air finds before circulating around the sidepod; thus, it must be designed to fulfill precisely this function: "start" to adjust the air so that it adheres to the sidepods and go where it has to go.

There is another similar device on the outside of the sidepods, whose goal is the same; you can also take advantage of the support of the mirror for this:

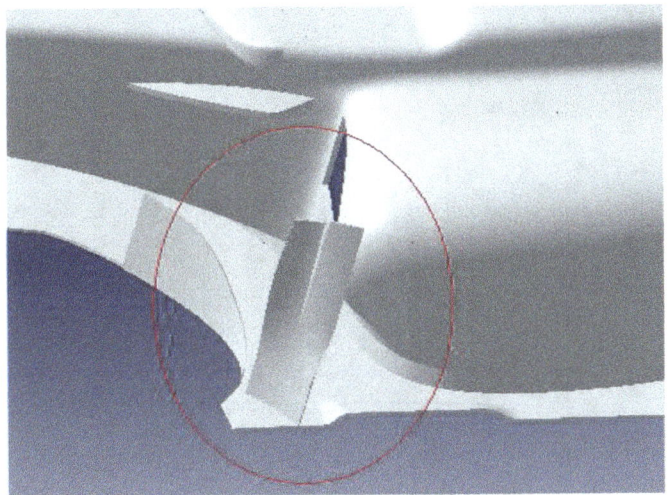

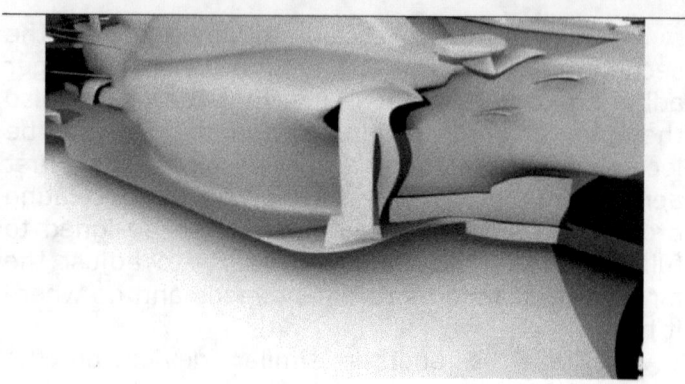

Another function of these parts, also called "Turning Vanes" is to reduce turbulence from the front wheels:

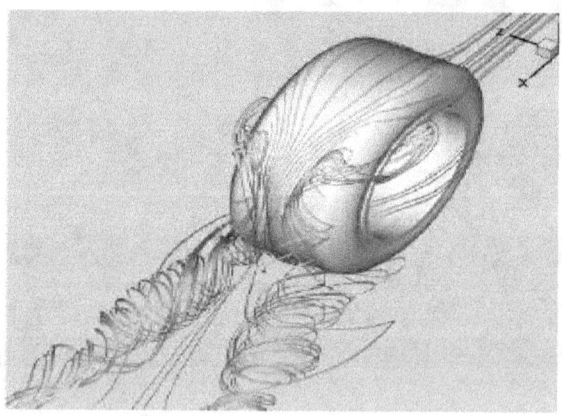

"DOUBLE" UPPER INLET: DRAG REDUCTION

It is essentially to reduce the overall drag of the car, reducing the low suction pressure that forms in the stern of the car; Mclaren and other teams are using "the turret" of admission, to place a "double" air inlet to redirect it towards the back.

This air that is collected, may or may not be coupled with air from the sidepods or air that cools the engine; thus all flow is used to fill said depression or low back pressure; the other destiny of this air can be the top of the diffuser to increase its efficiency as a "blown air":

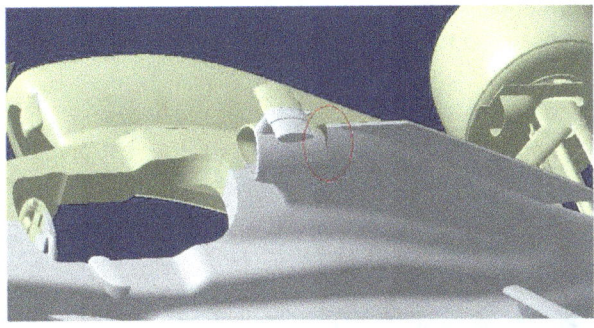

We saw at the time the "pressure releasers"; We can use that air from the high pressure zone to for example fill stern depression and thus reduce drag; i.e. depression used to suck air and move where needed:

AIR EXTRACTORS: INCREASING REAR WING EFFICIENCY

The main objective is to make the entire rear wing to function optimally:

They are susceptible of being placed in the bottom or as in the previous picture on the sides; the function is exactly the same as that already seen of the Gurney flap placed in the same place; in this case, a low outside pressure generated by the car speed (Venturi effect), producing an air suction inside the "box" of the rear wing; It makes it work better because all the air passing through the box is "useful" and does the job:

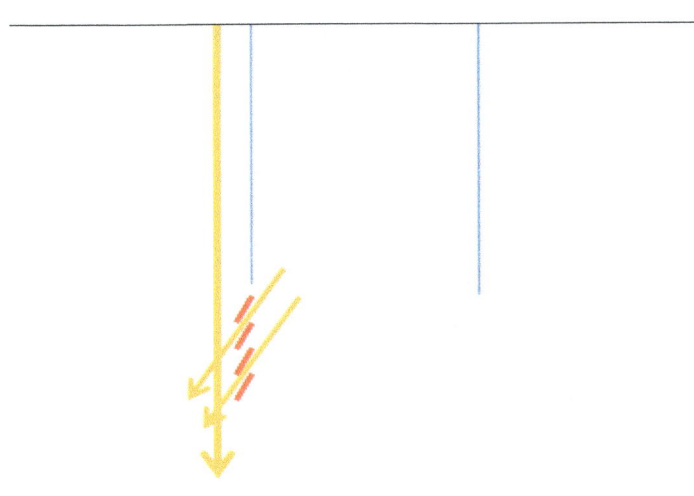

WARMING THE TIRES

The teams which tires suffer from a shortage of heating, either by the driving style of the driver or because the suspension has a geometry that causes little warm them, must manage to meet these deficiencies.

Use the brakes cooling gases to hit on the rim and this makes the tires become hot; directly heating the tires can also be another option.

This is typically used in the front because the rear can leverage the exhaust gases directly. Michelin has investigated the matter so heating only certain area of the rear tires.

Another effective and feasible method of heating the tires is to design a way that braking makes heat them; we are referring to design the process and method of braking regardless of driver, not the braking system.

LOWER CHANNELS OF THE FERRARI NOSE

Ferrari in the 2012 season incorporated several channels below the nose, which accelerate the air inside it causing a depression:

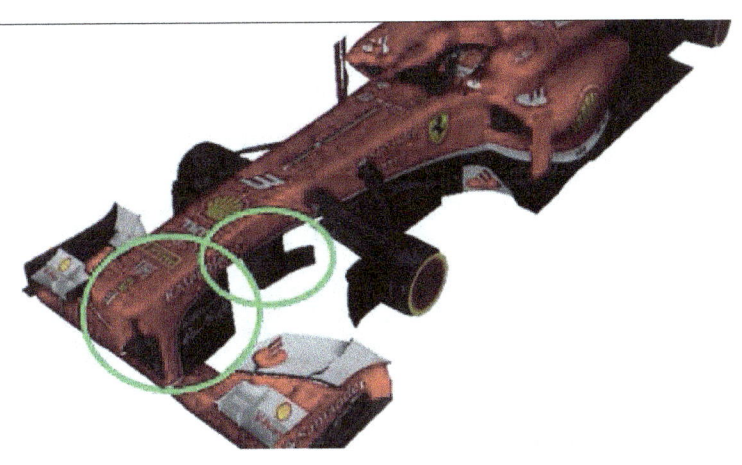

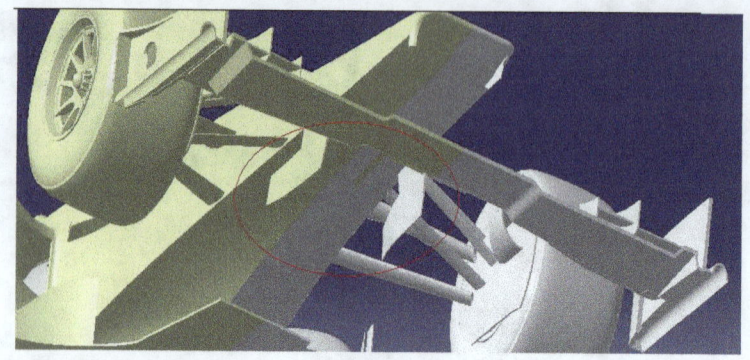

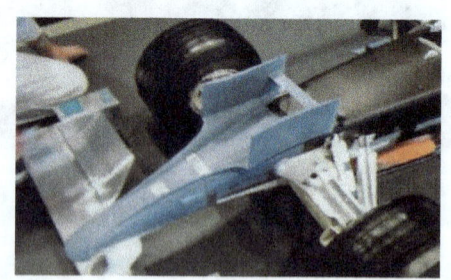

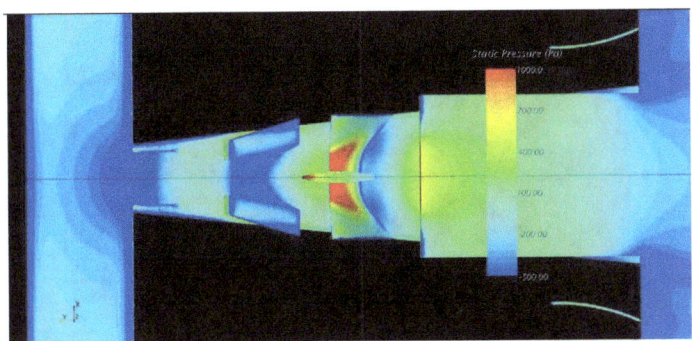

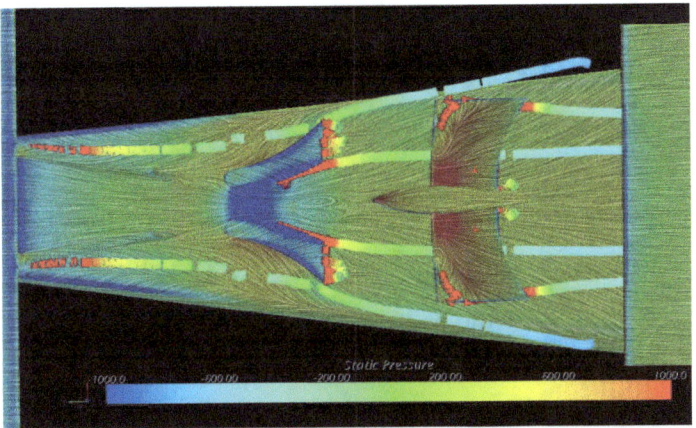

In the 2014 season we witnessed several similar ideas with the same goal: to create channels in the nose to increase speed, lower pressure and thus increase the load (also to help the Y250 vortex):

WING FLECTION

There are 3 cases:

- Full front wing deflection: If the wing flexes down, the benefits in terms of increased downforce are spectacular; The following picture shows the Red Bull wing is much lower compared to Mclaren:

- Deflection of a plane of the wing: we know that the more speed exists through the opening between the different levels that make up a spoiler, more angle of incidence can reach and therefore more downforce; We can therefore vary the "gap" between the

planes.
- In previous seasons we have seen attempts to "play" with the regulations, strict compliance with the limits of deflection marked: if we take the deflection of the ailerons, we can to reduce the gap between planes at high speed, causing the wing not produce a lot of downforce, then it is able to reduce drag and increase top speed (automatic DRS).

Most teams who introduced this concept made the piece to deform was the front wing flap; Instead, Ferrari made the main plane was which deflect, being very "obvious" the introduction of this system as an attempt to circumvent the legislation.

These deflections and vibrations must be taken into account, for inclusion in its case in the car's Post Rig analysis.

We also know that the car not only moves or vibrates up and down, but also makes it laterally, producing, with the heave vibration, a movement in 3D; to try to mitigate this lateral vibration, to image and similarity of the already seen vertically, you can use the same front wing to act as a mass damper but this time laterally; the combination of both artificial "mass dampers" would make the car, despite the vibrations, maintain a posture or "uniform" action on the asphalt.

VERY HIGH SUSPENSION OF FERRARI 2012

Ferrari, among other problems, has excessive tire wear. To improve aerodynamic efficiency of the car, it has decided to lift the anchor points on the chassis of the front suspension:

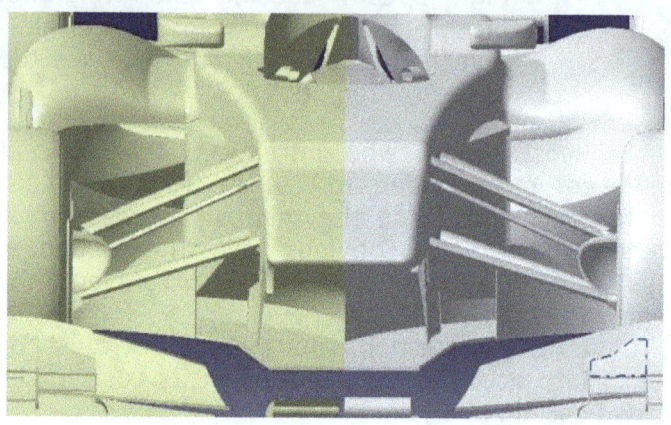

This kinematics of the front suspension means that when the wheels up and down, the distance between the front wheels vary excessively; this makes the tire rub much on the asphalt producing increased wear.

Furthermore, another problem is associated to this geometric arrangement of the suspension which is predisposition to having acquaplanning.

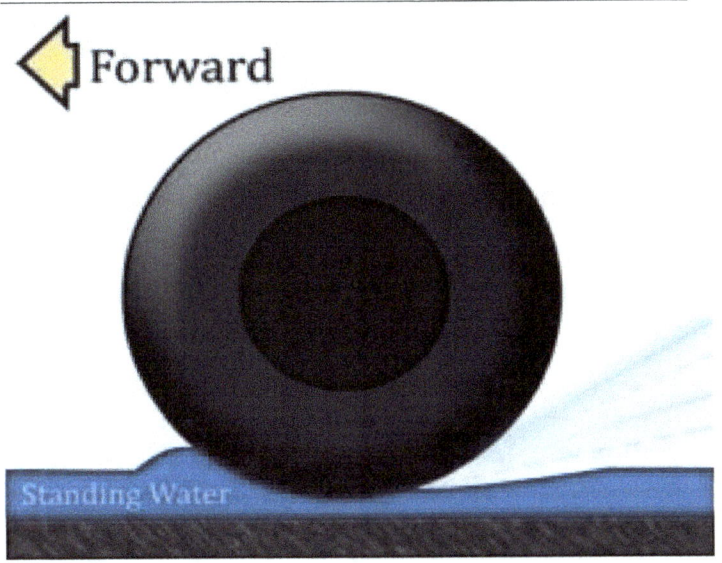

In the 2013 Barcelona Grand Prix, we have observed that Ferrari and Mclaren worked on the front suspension to prevent the change of the track due to the large angle that have respect to the "Y axis" which made them vary greatly front track of their cars, creating variation of mechanical grip of the front axle according to the variation of the compressive force in the front suspension by impact, braking, or by varying downforce.

There were many problems with the tires, which could not withstand the abrasive pavement of the Barcelona track, so only Ferrari with Alonso, understanding this, decided to run 60 laps at qualifying pace, using the largest number of new tires, and it was not problem of cars, but the softer Pirelli compounds used in Barcelona could not withstand the aggressive pavement.

Only they understood that it was not a matter of set-up, but it was a matter of the quality of compound.

It was a great action and took all northern Europeans off guard.

DRS: ANOTHER MOVEMENT CONCEPT

The essence of the DRS is to move the rear wing flap in a straight to have less drag, and therefore more top speed in order to be able to overtake.
Enrique Scalabroni variation is to keep fixed the flap and move the rest of the wing; We've tried getting less drag than with the previous system, with the same downforce.

The rules of the use of the DRS is clear: if enabled, there must be a distance between the profiles of 50 mm:

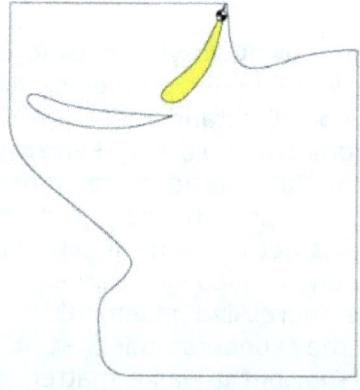

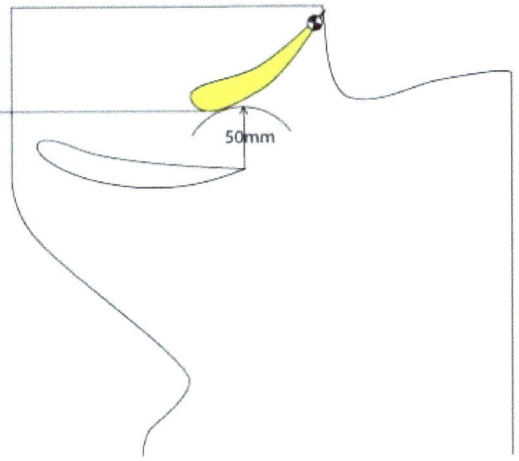

With the new optimized version:
On/off:

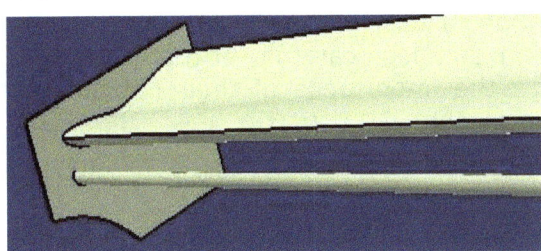

We have already done CFD tests with this system and it works perfectly; maybe we see it soon on a F1 car....

The concept or idea of the DRS has brought a revolution in terms of possibility and number of overtaking. Just look carefully at the following chart to realize the importance of their implementation and development:

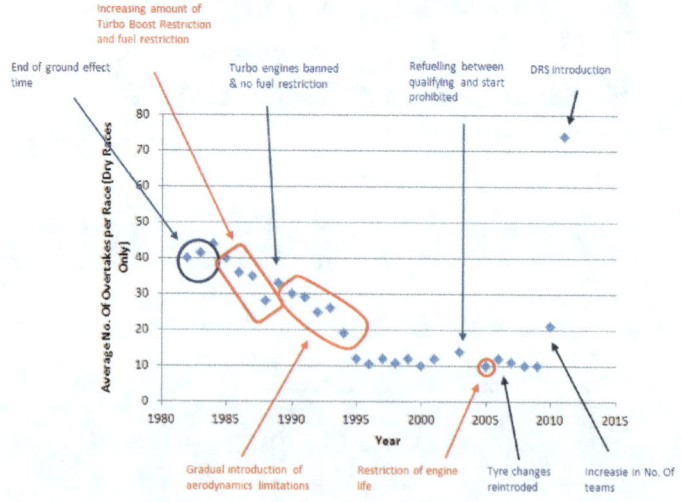

RED BULL FRONT WING DEFLECTION BULL; AERODYNAMIC AND SUSPENSION IMPROVEMENTS ?

At the end of the 2012 season we saw quite clearly the deflection of the front wing of Red Bull.

The obvious question is whether it is legal? Yes; they passed the appropriate tests by the FIA and passing them, they are legal; nothing more to add to it (also not the aim of this book, as we know).

The other question to ask is whether this deflection, which is "obvious", can be utilized for other things.

We all know the mass damper; We do a thorough study of it in the Post Rig Analysis section.

This is definitely a top-down oscillating weight that was placed inside the nose of Renault, to improve the grip of the front tires; it provided many improvements from a dynamic point of view and grip.

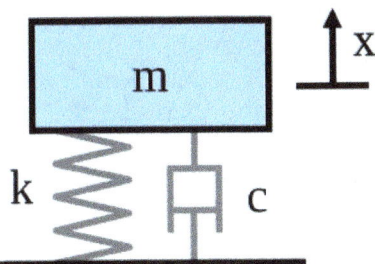

Mass Damper exists not only in cars, also in buildings with the same goal: in this case, mitigate earthquake vibrations:

Red Bull take advantage "legally" of the vibrations of the front wing, to get the same effect as Renault Mass Damper.

The original mass damper had quite mass; this time, being the wing more advanced, is supposed to have less mass to achieve the same (has higher lever arm or torque). The question is precisely taking advantage of the vibration and the rules for their own benefit....

Therefore, Red Bull keeps deflection limits set by legislation, though, it makes it vibrates at a certain frequency in order to mitigate the rebounds of the suspension and increase the grip of the front tires.

To calculate the required mass of the mass damper:

- The first is to determine the frequency of interest you want to mitigate (Post Rig analysis or track).
- Mass Damper parameters is then calculated using the following formula:

$$f_d = \frac{1}{2\pi} \sqrt{\frac{K_d}{m_d}}$$

Of course you want to have the least possible mass to not affect the center of gravity, but the smaller the mass of the mass damper the lower its effect, so you have to reach a compromise.

For the Red Bull nose+wing the mass is already imposed and therefore have to be calculated only the spring ad damper.

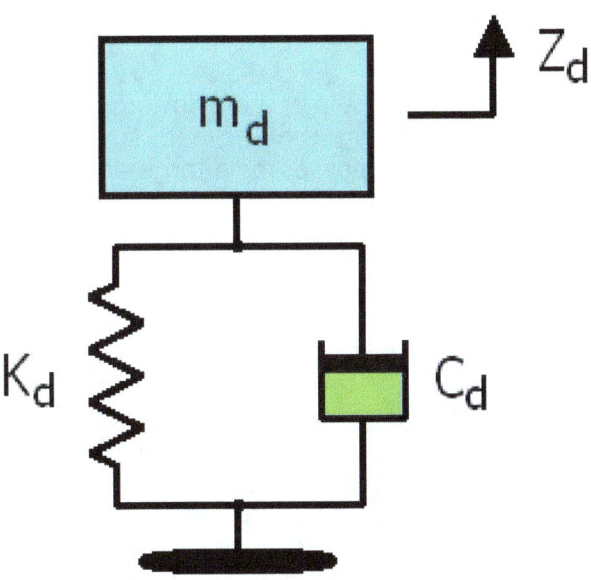

The nice thing about using the nose + wing as mass damper is that being so far from the center of gravity, the mass must not be too large to be noticed the effect.

By deflecting the front wing we can vary the downforce generated; i.e.: we can generate more downforce if we properly combine the aerodynamic vibrations, vibration suspension and vibration-deflection of the wing; this will be discussed later in the section aero-post-rig.

INCREASING DIFFUSER EFFICIENCY

If we use Naca intakes we may be able to increase the efficiency of a diffuser. Simply place them in the arrangement shown in the following picture.

The intakes help the air flowing through the bottom to run along the inclined surface of the diffuser although it possesses a large angle:

"VISUAL" AERODYNAMICS; ANALYSIS

We are sure that as lovers of the technique of competition, leaf through countless times technical nature websites to find the latest breakthrough. We are also confident that most sites explain the new systems as opposed to its "no" existence.

As engineers, we have asked the question of why exists and how certain system functions; We can answer us observing the context in which it is installed but often it can be difficult to properly set the operation:

We can have ideas about their role and be all true, all false or half....

Obviously it is a visual element analysis, but barely explain anything what it does and what its usefulness is.

Thus it becomes very useful, as we have done ourselves in this book, to speculate about the function and analyze each of these possibilities.

An example: we see the following Williams and Red Bull wings with DRS enabled; the photos were taken directly from the TV:

We can see that the 2 openings are different, being Red Bull's much larger; the upper flap is substantially flat, whereas the Williams not; drag of the Red Bull will be much smaller.

Apparently it's like Red Bull vary more degrees the movement of his DRS: By regulations all rear wings must move the same amount; then: why these pictures tend to say that it not?
Because the initial angle or position of the Red Bull rear wing is larger; so, when activated, it appears to have moved further.

The origin of this fact is also simple and indicates the immense possibilities that the Red Bull had (2011): He is able to get out on track with the least wing load generated by the rear wing of the whole grid, because they have left downforce.

Incredible capacity that had the Red Bull; he had more than enough downforce. And he had more top speed. They can afford to have very little drag (wing virtually flat) to activate the DRS:

In the next picture we appreciate precisely this difference in ability to place "more pitch":

SELF-GENERATING SYSTEM

We know that a depression zone tends to suck and attract air for filling said depression; as we have seen on several occasions, systems and locations.

We can also use this fact to communicate a high pressure zone with another low pressure to uniformize both areas.

Let us imagine the following scenario:

We put a tube that connects the nose with the back of the car: In front we have high pressure, while the rear there will be low pressure:

As the car has more and more speed, the pressure will be higher front and rear depression will also be greater; this is just what is called self-supporting system: the more speed, the system sucks more air.

The same principle was used by the tunnel nose by Ferrari: at the top had depression that sucked air from the bottom.

Here is another very clear example:

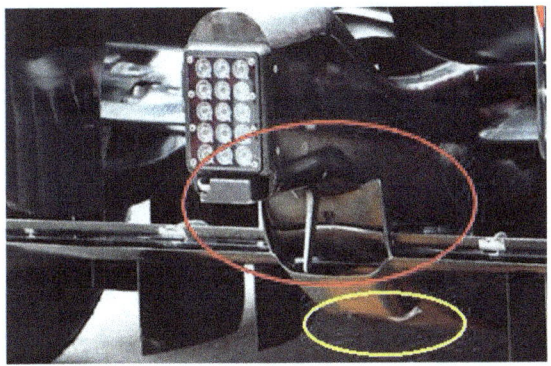

In the rear depression exists; air is sucked through conduit whose start is the clear circle which in turn is an area of overpressure and air is ejected through the top opening (dark circle).
All this system acts as small double diffuser to increase efficiency.

TRUCKS PRESSURE RELEASERS

We've all seen the small openings that have the most of the trucks in the front and side.
The goal they have is to relieve pressure that focuses just on the front by direct impact of air and moving the extra flow of air (not used in cooling radiators) to the side; it reduces greatly the total drag of the truck; This system is a self-stable or self-generating system.

More or less, are have the same effect than the "craks" in the endplates in the rear wing.

INCREASING DOWNFORCE UNDER THE CAR

We already know the channel placed at the floor:

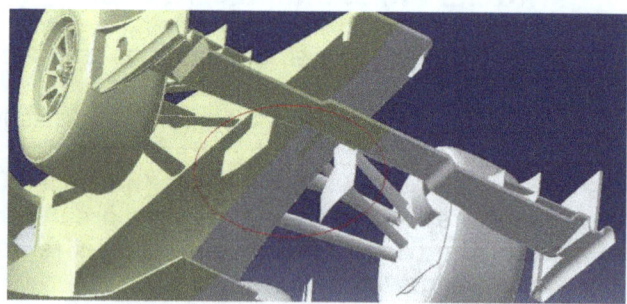

Ferrari has gone beyond the optimization of this system: tries to increase the area of expansion of air in order to increase depression and increase the suction under the car.

As we already know perfectly well, if we increase this excessively opening or expansion, the airflow can be removed from the surface making the system ineffective; to do this, we can apply the same principle used in their time to increase downforce of a profile if it had a large angle of incidence:

Separate it into parts.

In each opening a drop in pressure that sucks air and prevents the airflow running through within the "channel" to separate from the walls occurs. In this way, we are able to increase the area of expansion:

Another of the objectives that this further expansion can have is to better adapt the air leaving from other areas; for example the Y250 vortex.

IMPROVED DIFFUSER PERFORMANCE: SUPPORTING COANDA

We already know that to optimize a diffuser, just try to generate a depression over the diffuser so that this vacuum suck the air passing through the diffuser:

Zona de baja
presión

Ferrari engineers, probably by the need to optimize and improve the car, came up with a brilliant idea:

This is taking "extra" air, not from the air circulating around the car and use it to channel it toward the top of the diffuser; Where do they take it?

From the cooling flow that goes through sidepods itself.

They have prepared a series of flakes on the bottom of the sidepods so that the outside air due to its high speed, suck air from inside incorporating it into the mainstream of air going into the rear.

As if that were not enough, they go further:

Why they take air from said area and for example not higher?

The answer is simple:

To take advantage of the longitudinal and stuck to the surface of the sidepod component; thus added "Coanda effect" to the mainstream.

It's a great idea as they take advantage of flow or quantity of air that otherwise would be wasted downstream. In our opinion, there are 2 types of ideas:

- The great ones, as this one from Ferrari, and others such as blown diffuser, or the double diffuser, etc.
- The very elaborate ideas for example changes in a front spoiler.

This does not mean that great ideas have not working; Take for example the case of flakes or gills of the sidepods:

- You need to know where to place the gills: height.
- You need to know where to place the

gills: longitudinal position.

- You need to know how many flakes are optimal.
- You need to know what flakes size is optimal.
- Etc.

Each of these needs involves a very complicated study that either can be done by using wind tunnel or CFD and simulation techniques; optimal, as always, is an appropriate combination of both methods.

SUPER EFFICIENT DIFUSSER: MODIFICATION

We will try to modify the diffuser F1 to increase efficiency; we will not consider any legislation; it is simply a design exercise.

What we pursue is to increase the angle of the diffuser as well as their area so they get passing a large amount of air under the car and increase downforce generated by the floor.

We know what the diffuser is and especially what the problems associated with increasing its output angle are:

The main problem is that the air flow tends to detach from the surface of the diffuser if the angle is too high.

Now consider a wing; we have seen that one of the ways to achieve greater angle of incidence in a wing is to separate in pieces:

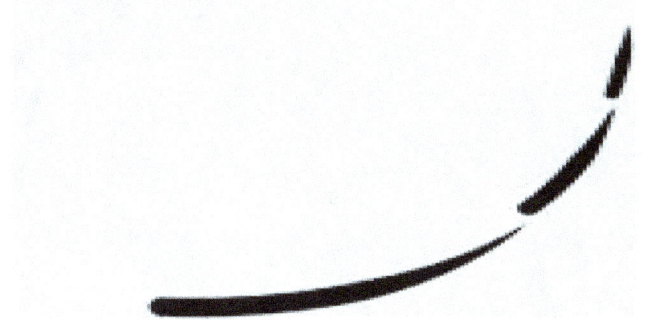

We do exactly the same in our diffuser; ie several planes placed above the main diffuser; in this way, we can increase its angle of incidence. The current flap diffuser works the same way.

The problem appears when the flap diffusers now included receive airflow: we must allow plenty of air from the front to get there and in good condition; it not is easy, really.

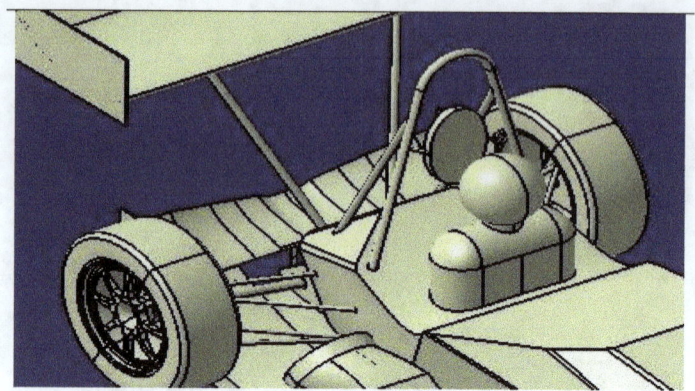

Look at this old model of Formula SAE: it is virtually impossible to reach any amount of air in "good conditions" to the flap diffuser. There are many previous interference.

GROUND EFFECT OPTIMIZATION: EXTRACTION FLAKES

One of the fundamental and most important objective in competition is to maintain the low pressure at the bottom between the asphalt and the floor of the car; in this way we will be producing a large amount of downforce. To maintain it we already saw various methods and systems; this time we see the following picture:

There are several flakes in the direction of movement: the higher the car speed the greater the air extraction will produce right in the outside of the floor.

Because of the creation of low pressure, the surrounding air will want to enter said depression, but then, the flakes will suck either the air that tends to introduce or has already been introduced.

Therefore we generated "artificially" a kind of lateral "skirt" in order to prevent penetration. Really great idea.

The faster the car go, the downforce it will have; This is a typical problem of the systems we call "self-generating". This result is detrimental since it will generate much downforce in a straight when not needed. It is a passive system which should be put some sort of automation as limiting the downforce depending on speed; ie that at a certain speed, not produces more downforce.

As we saw, one of the methods used to not allow the entry of air at the bottom, is the use of "artificial" side screens; this is the case of using exhaust on the side:

DOWNFORCE INCREASE AND REDUCTION OF DRAG; AUTOMATIC METHOD

After seeing all that we have seen and analyzed, we know for sure, that the dream of every aerodynamicist is to increase downforce in cornering and reduce it in a straight; What's more, ideally this regulation would be automatically done, without any intervention of the driver; this last consideration is given for reasons of technical regulations.

In the 2013 season, Team Lotus incorporated an amazing system which in our opinion is an absolutely brilliant idea, the kind that will be remembered for years. Aerodynamically speaking, it is fabulous: a fully automatic system, which regulates the downforce depending on whether the car is straight or is curved.

How does it do that? Based on the car speed; in a straight the car is much faster than in corners; it uses precisely this, to self-regulate the amount of downforce.

One of the most important pieces in amount of downforce is the rear wing and one of the areas that produces more drag is the rear of the car.

Lotus worked on these two areas: on the rear wing to increase downforce and on the stern to reduce the overall drag of the car. How? Here is the crux of the problem.

The car has air inlets at the sides of the intake port; This air is channeled into two sites:

- The rear wing and injecting so as to increase the downforce.
- Towards the stern, in order to reduce depression.

The trick lies upon which the regulation of force is the tube that carries the air:

- When the car has a lot of speed the flow of air comes out the rear, reducing drag and thus increasing the top speed.
- When the car has low speed (curve) airflow exits through the rear wing, increasing downforce.

Absolutely brilliant; It is a great exercise to design the duct to get what Lotus achieved; think that the Coanda effect have much to do .... Perhaps we can also install a nozzle to form a sonic plug...

Magnificent example to think, design and test it using CFD techniques.

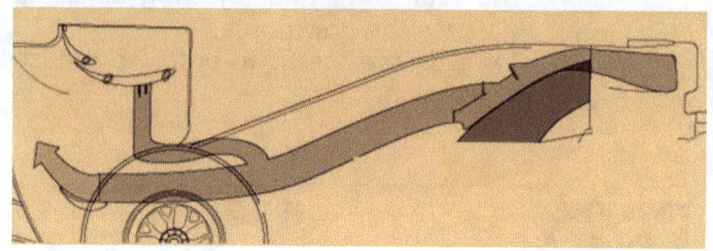

Why no air is introduced into the floor to reduce downforce? Because the floor barely produce drag.

Mercedes incorporated the same system:

And the intakes for channeling the flow of air are others:

To reduce drag air is injected into stern depression; the tube that carries the air finishes in a kind of diffuser:

How this diffuser work and why has this peculiar geometry ¿?

The stern area where there is the "Great Depression" is a large area distributed on the entire rear area:

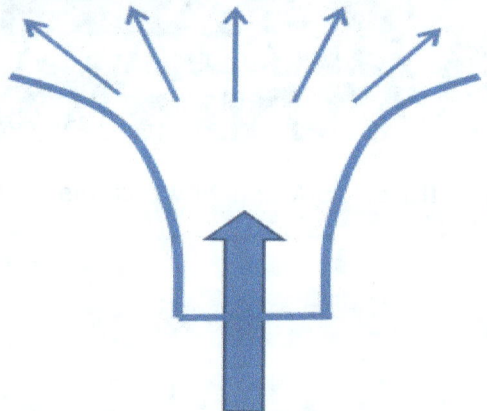

Thanks to this form of "diffuser" or trumpet, adequately diffuses the air to the entire area of depression, reducing it in its entirety; also it takes advantage of the Coanda effect to stick to the surface defined by the trumpet.

We may combine the previously seen precipitator system to better diffuse the air coming out of the exhausts or cooling?

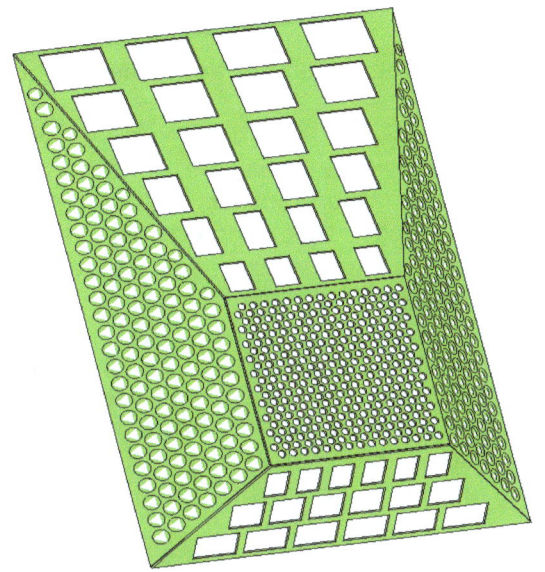

DRAG REDUCTION, REDUCING THE THICKNESS OF BOUNDARY LAYER: FERRARI LOWER INTAKE

We saw earlier what was the boundary layer, its development, its consequences and the methods that exists for their control. Its control was essentially based on:

- Control its thickness.
- Control its adherence.

The more "fat" is the boundary layer means that there is more air with less speed than the surrounding air; this means that the car is losing some of its energy in braking that amount of air; the car will have less energy available to achieve greater speed. It is about "energy" balance reasoning, but they are real and teach us what to do.

We also know that the thickness of the boundary layer, among other factors, is directly related to the length that runs through the air on a surface; that is, the larger a surface, the thicker the boundary layer.

If you look at the nose of an F1, we see that the length between the tip of the nose and keel at the bottom is quite long; it stands to reason that under the driver more or less the thickness of the boundary layer is large.

To eliminate this great thickness and reduce it, Ferrari with the 138 model, introduced a lower opening:

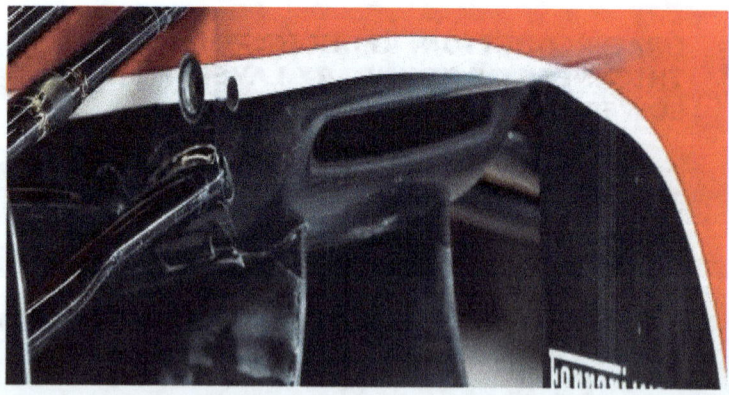

This opening or "mouth" sucks the boundary layer that is formed ahead significantly reducing the overall drag of the car; we must take into account the reduction of the amount of air going to circulate under the ground now ....

Red Bull already did something similar, although the "extra" function was to reduce the pressure of impact in this area, like Ferrari:

Now that we know the function of the "take-mouth", let's see where the air is introduced: Ferrari introduces air into the cockpit and thus take advantage of the depression generated in the cockpit to suck the flow through the "lower mouth". In the season of 2014, we saw Red Bull use something like this Ferrari system; Duct called "S-duct" because sucks air from the front and expels it through a slit in the top of the nose; the idea is a little variation from the studied now, because its aim is to delay the turbulent boundary layer over the nose:

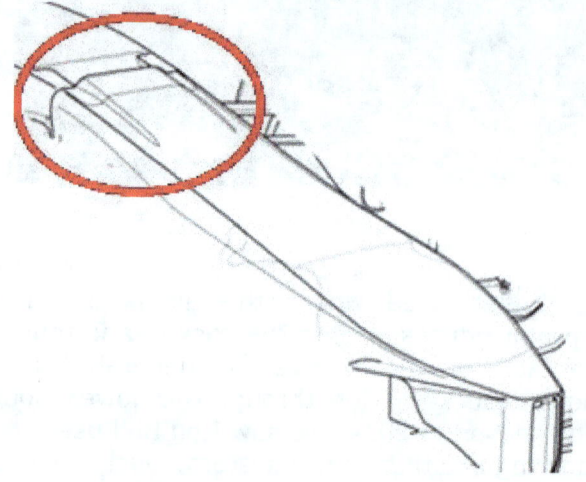

In our humble opinion, there is another alternative outlet airflow: located at the stern, as Mercedes and Lotus did with its passive drag reduction system. We made some CFD tests and the results say that the efficiency is much higher for several reasons:

- In the stern there is more depression and therefore more suction force.
- It is also a self-generating system.
- Air is injected "precisely" where more drag exists, the car's total drag being reduced appreciably.

We have read many articles on the internet saying that Red Bull was the first to incorporate this system and that Ferrari has improved; maybe it applied to race cars, not doubt; but aircraft used for almost a century ago; Consider the "BLS" seen above.

We have also seen a control system of the boundary layer in wind tunnels using "exactly" this method: the suction of the boundary layer.

"THERMAL" DIFFUSER

Just after the 2013 Grand Prix at the Nurburgring, began circulating around the paddock the rumor that Adrian Newey introduced a system in which the diffuser and surrounding areas vary their geometry in function of temperature....

Consider that right there is where it goes the engine exhaust gases and temperatures of 300 ° C are easily achieved; to modify the inclination or curvature of the diffuser greatly influence the amount of downforce that the floor can generate; great idea.

TIRES: 2013 SILVERSTONE EXPLOSIONS

Tires are an elemental piece, also from an aerodynamic standpoint, since the heights of the car with respect to the road are directly conditioned by its deformations. In the 2013 season, all teams discovered, at least until Silverstone, that the main trick for the Michelin tire to work optimally was to warm them; the team that don't get it over was lost ....

The structures of the tires to Silverstone GP were made of steel and this should work at higher temperatures than the tires with Kevlar structure or carbon fiber for example, so the teams had to put a set-up that makes raise the temperature of these; To achieve this, they had to rush and force operating conditions:

- Lowering the pressure below 16psi (1,10316bar).
- Placing aggressive camber incidence, higher than 2.5 °.
- Rotating tires (left to right) which meant that the tire would work in the opposite direction.

Swapping them, the material and structure were not prepared to work as they were forced to work, despite making good times on track.

This put more stress on the structure of tires; the disposal of the tires to work excessively until burst:

Pirelli decided to change the structure of the rear tires from steel to kevlar, and made the following recommendations for the Nurburgring GP:

- Standard Minimum pressure on the front and back: **16psi (1,10316bar).**
- Minimum pressure of rolling in the front: **20psi (1,37895bar).**
- Minimum pressure of rolling in the rear: **19psi (1,31000bar).**
- Maximum negative camber on the front axle: **4.0 °.**
- Maximum negative camber on the rear axle: **2.5 °.**
- The front and rear tires should be used on the side of the car for which they were designed: do not change a tire side.

Keep in mind that the structure of the front tires has not changed: still steel, which means that the front should work at a different temperature than the rear, must have a very different set-point to make them work properly.

Make them work properly, involves making a good study: aero - post - rigiiii.

TOP SPEED INCREASE USING INLET ADMISSION-1

When the car is in a straight, taking admission is attracting more air than required by the engine; precisely because of that the pressure front bubbles already seen that involve at the same time Hemmoltz reverberations are formed.

If we were able to use this air surplus to reduce drag, we would have an automatic "DRS".

To achieve the above, we can place a valve in the intake port or air box, so that it opens when the "DRS" is activated; thus, we can direct the air that is left over to the rear of the engine cover, to fill the stern depression and reduce drag.

Obviously we are losing "something" useful power, but the power loss was amply recovered by reducing drag and thus increase the maximum speed of the car in a straight.

DRS Abierto
Se abre el paso

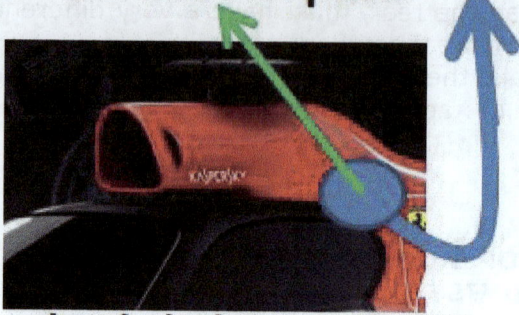

**Alta velocidad; el exceso de aire
en el air-box, se utiliza para
reducir la drag.**

In turn, with the "DRS" closed, the valve is closed
so that there is no air outlet:

DRS Cerrado

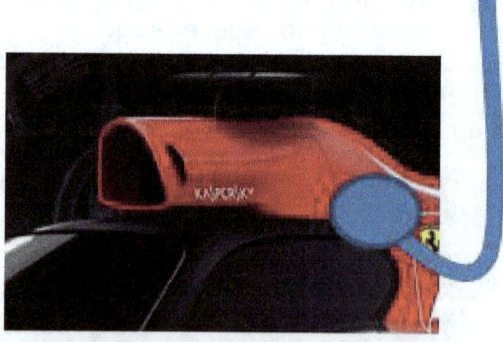

Baja velocidad.

TOP SPEED INCREASE USING INLET ADMISSION-2

It is a marriage between the system discussed in point 41 and point 45.
At high speeds, the excess air goes out of the air box (without valve) towards the rear taking care to reduce drag; great new idea; absolutely brilliant.

INCREASE DIFFUSER EFFICIENCY THROUGH THE SUSPENSION ARMS

In the 2014 season, the FIA banned the "Wing Beam" or wing-diffuser; It entails, as mentioned, an incredibly reduction in performance or efficiency of the diffuser, and therefore, a very significant reduction of the rear downforce but also the total.
The McLaren team has developed an innovative and incredible system that is based on the rear suspension arms:
The arms are wide:

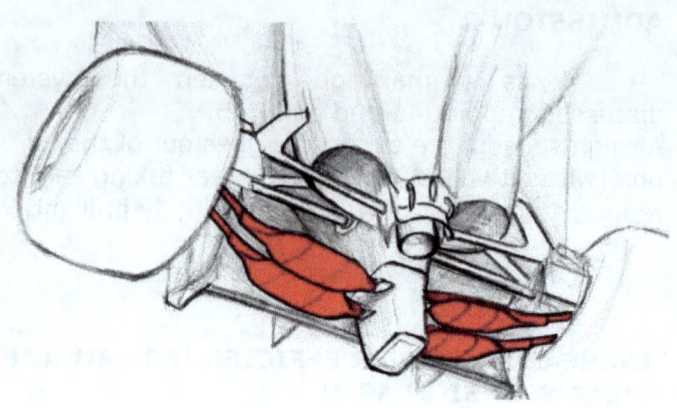

The arms separate and then join according to the aerodynamic load; It entails air to pass or not pass between the arms; if we are able to divert air up, we'll be simulating "artificially" the wing beam.

Absolutely fantastic system, worthy to stand in the books as an amazing novelty. Why is correct by rules ? because the rules say that the arms profile must to be simetric....

But how does it work? The essence lies in the profile shape of the arms; Another extremely important issue is the relative position of both arms and their separations:

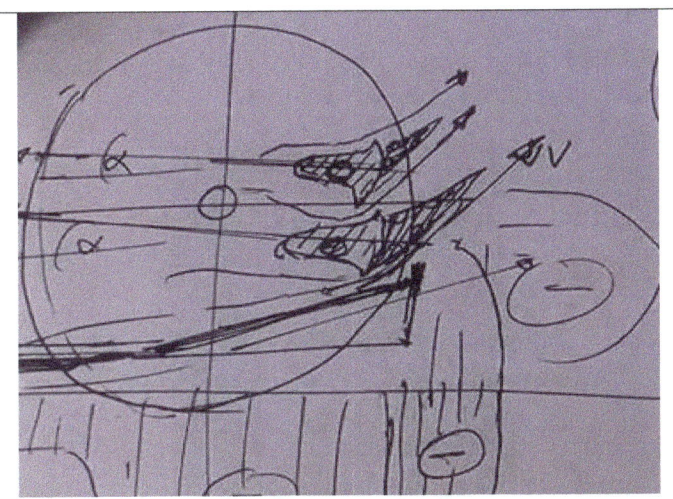

This diagram is performed by Enrique Scalabroni, and perfectly illustrates **the arms' profile**.

SMALL APPLICATION TO CYCLING

We already know the effect of the turbulators and we know various applications of them; but we can go a step further.

Let us imagine cyclists' legs:

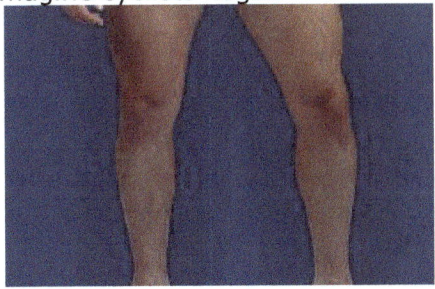

We can only shave the back of the leg leaving hair in front; thus, this hair, acts as turbulator; effective method, we can ensure:

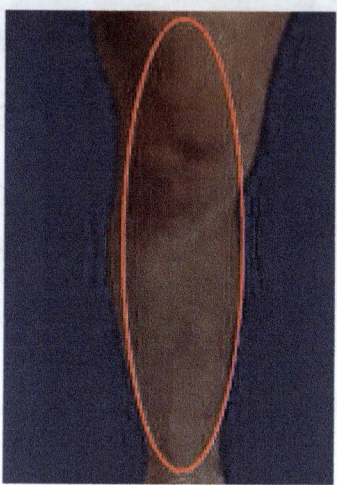

DRAG REDUCTION AND TOP SPEED INCREASE

We can think of the following method:
Opening over the nose of the car using Naca intakes or similar; said intake collects air channeling to the sides of the pilot to the rear of the car.

We can also think of a system that takes air only beyond a certain speed.

This system would be more efficient than traditional DRS, because we aspire lot of air and injected into the rear area of low pressure, right where most drag occurs.

2014 MERCEDES ENGINE: GREAT IDEA FOR EXTRAORDINARY POWER DIFFERENCE

We have already known the power Mercedes engine had in 2014 season; its difference with the other cars could be close to 100 horsepower.

What is due this? Let's look and ensure the very close dependence between aerodynamics, cooling and weight distribution.

We know that the motor is turbo; the turbo system is rotated by the exhaust gases and rotates a compressor that compresses the air going into the pistons; This compressed air causes the engine has increased power; This is the essence of the turbo.

The turbo-compressor arrangement is:

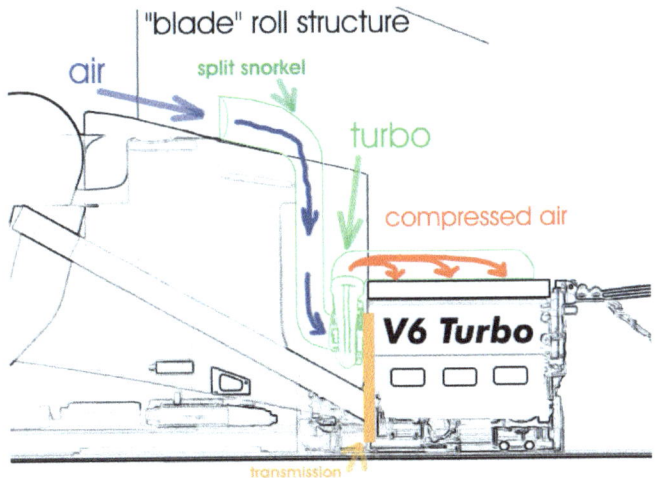

The turbo is placed right in behind the exhaust outlet; the compressor is next to the turbo:

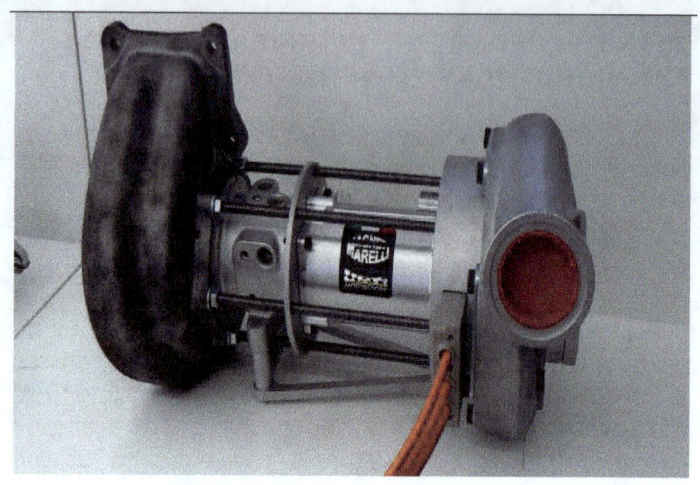

The turbo is heated greatly, logical, and it makes the compressor to be heated too; the air passing through the compressor, although compressed, is also hot, with what is necessary for cooling so also the engine power is increased; this is done by the called intercooler:

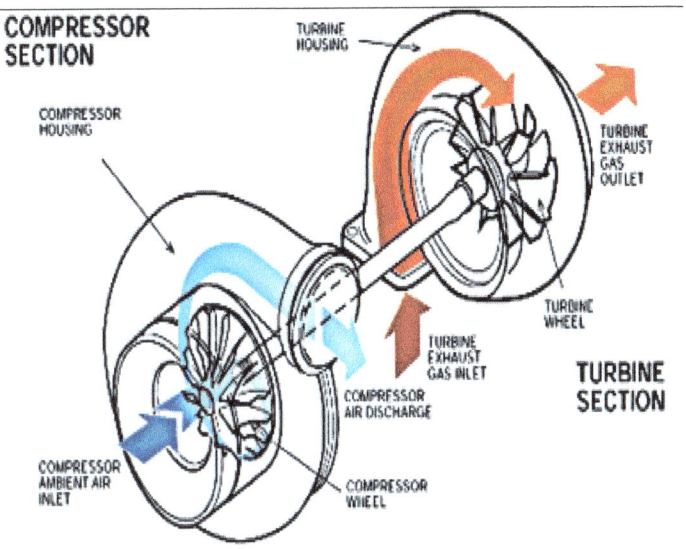

COMPRESSOR SECTION

COMPRESSOR HOUSING

TURBINE HOUSING

TURBINE EXHAUST GAS OUTLET

TURBINE WHEEL

TURBINE EXHAUST GAS INLET

COMPRESSOR AIR DISCHARGE

TURBINE SECTION

COMPRESSOR AMBIENT AIR INLET

COMPRESSOR WHEEL

Mercedes had the great idea:

This is to separate turbo and compressor: the compressor is placed in the front of the engine mated with the turbo with a shaft rotating it; thus the air compressed by the compressor is cold; therefore it is not necessary to have an intercooler or at least have a very small version; with all this:

- It reduces or eliminates the intercooler.
- The sidepods need not be as great, since it is not necessary to cool the intercooler.
- Being the sidepods smaller, you can reduce them in order to reduce drag and have more top speed.
- By not placing the compressor behind, there is more space to play with weights, for example the gearbox, so you can dynamically balance the car better.

LIMIT OF POWER TRANSMISSION

Introduction:

When we want our car has more acceleration we always think about increasing engine power, and rarely think about changing our tires. In this section we intend to show that power is not everything.

It is important, very important, but it has its limits; and it is given by the friction coefficient μ of the tire.

We introduce the car:

Mass		
Mass	1690	kg
Wheelbase	2.731	m
Weight split (% front)	50	%
CG Height	0.5	m

Conditions		
friccion coeff	1.1	-

Transmission		
Primary drive	1	:1
1st	4.23	:1
2nd	2.53	:1
3rd	1.67	:1
4th	1.23	:1
5th	1	:1
6th	0.83	:1
Final drive	3.63	:1
Wheel R	0.33	m

Engine		
Speed (RPM)	Torque (Nm)	Power (Cv)
1000	341	48.5
1500	387.5	82.7
2000	418.5	119.1
2500	449.5	159.9
3000	465	198.5
3500	480.5	239.3
4000	511.5	291.1
4500	542.5	347.3
5000	558	397.0
5500	542.5	424.5
6000	527	449.9
6500	511.5	473.1
7000	496	494.0
7500	480.5	512.7
8000	472.75	538.1
8500	403	487.4

We begin varying the torque percentage to find the limit for which the acceleration of 0-100 km / h is made constant; is the maximum that is possible with this car and these tires.

Results:

ORIGINAL μ (1,1)				
0-100km/h	5.66 sec	0-? Km/h		120 (km/h)
0-160km/h	12.52 sec			7.64 (sec)
0-200km/h	20.40 sec	Potencia		347 CV
0-400m	13.86 sec			
0-1000m	24.94 sec			
top speed	279 km/h			

20% Mas μ(1.1)				
0-100km/h	5.26 sec	0-? Km/h		120 (km/h)
0-160km/h	10.83 sec			6.91 (sec)
0-200km/h	16.90 sec	Potencia	416.4 CV	
0-400m	13.34 sec			
0-1000m	23.73 sec			
top speed	295.2 km/h			

40% Mas μ(1.1)			
0-100km/h	4.99 sec	0-? Km/h	120 (km/h)
0-160km/h	9.69 sec		6.42 (sec)
0-200km/h	14.64 sec	Potencia	485.8 CV
0-400m	12.93 sec		
0-1000m	22.77 sec		
top speed	316.8 km/h		

55% Mas μ(1.1)			
0-100km/h	4.92 sec	0-? Km/h	120 (km/h)
0-160km/h	9.12 sec		6.21 (sec)
0-200km/h	13.48 sec	Potencia	538 CV
0-400m	12.71 sec		
0-1000m	22.22 sec		
top speed	331.2 km/h		

60% Mas μ(1.1)			
0-100km/h	4.92 sec	0-? Km/h	120 (km/h)
0-160km/h	8.98 sec		6.18 (sec)
0-200km/h	13.18 sec	Potencia	555 CV
0-400m	12.65 sec		
0-1000m	22.07 sec		
top speed	333 km/h		

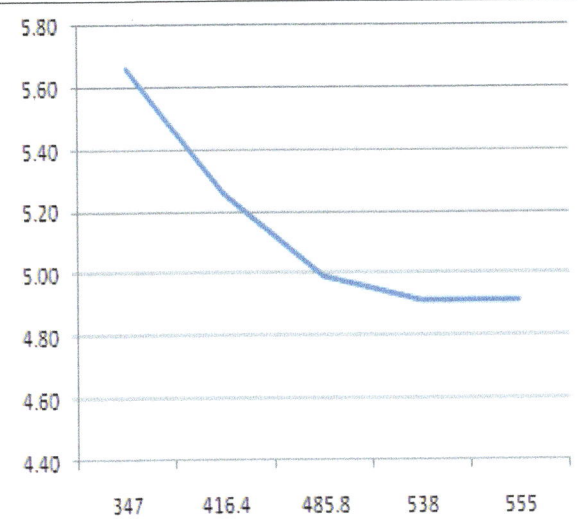

We see that the limit given by the tire is at 55% increase in torque. This can be remedied by placing a softer compound tires, or what is the same, increasing μ thereof.

Now we present the result with tires 10% softer:

55% Mas μ(1.2)			
0-100km/h	4.49 sec	0-? Km/h	120 (km/h)
0-160km/h	8.70 sec		5.79 (sec)
0-200km/h	13.06 sec	Potencia	538 CV
0-400m	12.38 sec		
0-1000m	21.87 sec		
top speed	331.2 km/h		

PRESSURES AND TEMPERATURES OF THE TIRES

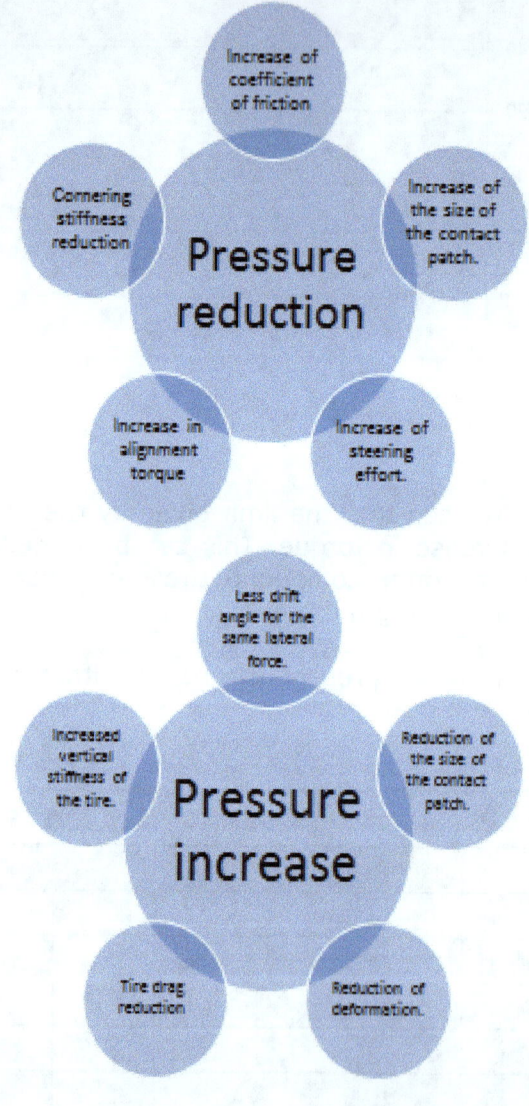

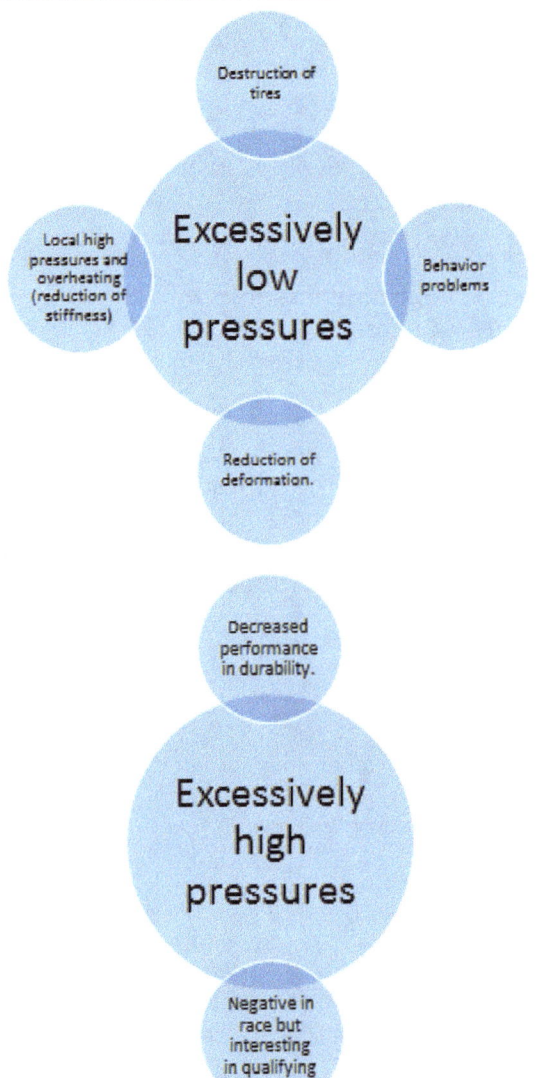

The temperature can be measured:

- On the surface of the tread of the tire by infrared sensors.

- Inside the tread or casing by pyrometer needle.

The sensor needle is inserted into the tire in three points:

- Extreme interior.
- Center.
- Extreme exterior.

This measure allows to know the way that the rubber works on the contact surface.

Changes in temperature will change the elastic modulus of the rubber and affect cornering stiffness.

Excessively low temperatures	• The rubber does not reach its optimal operating point. • Poor adhesion.
Excessively high temperatures	• Tire destruction by creating blisters.

Each tire has its optimum temperature for maximum grip: 80 ° C - 100 ° C (measured in the box).

BECAUSE MICHAEL SCHUMACHER DID NOT WIN IN HIS RETURN TO F1?

One of the theories that attempt to justify the poor performance of Michael Schumacher (MS) in 2010 and especially about his teammate Nico Rosberg (NR) is referenced to the dynamics of his Mercedes.

This article will discuss this behavior with the help of some images on board of the team Mercedes GP and more precisely on the MGP W01.

When MS came to Mercedes found a formula developed by his former driver Jenson Button during the years of Honda and Brawn. A car adapted to the fine style of management of the British driver. For those who are more accustomed to technical terms one could say that the car had an understeer tendency.

Add to all this the conservative front Bridgestone performance of 2010, which complicated so much to other drivers such as Felipe Massa, which enhances the tendency to lose the nose in the course of the curve.

What is an understeering car? When the front axle loses grip, the car turns less than indicated by the steering wheel and the front end tends to slide off the curve on a tangent. This trend is primarily determined by the weight distribution between the axles and the inertia that mass transfers produce during cornering. Anecdotally, the ex-driver and rally world champion Walter Röhrl once said: "Understeer is when you see the tree against which you are going to crash. Oversteer is when you only feel it.".

Those who have read about MS, know his driving style could be described as "aggressive" especially in the first moment you start to rotate. He likes a chassis that has a pronounced nose. That is, with oversteer tendency. Contrary to the "soft" steering wheel control of the above named JB.

In the words of Schumacher himself: *"In Mercedes, the pilot who was here before me, had a very clear style, and asked the team that the car had certain characteristics. I have not come to that. I can drive a car named understeery, but I cannot drive it as fast as the fastest car".*

We can also extract words of the engineers who worked together with the German to join this prelude of information: *"Michael believes that the duty of the car is to put the front wheels where he wants, then he will take care of the rest of the car to follow to the front wheels without losing control".*

We want to support what we discussed with images from the on board cameras of the Mercedes GP team, observe the behavior of the two formulas.
Here we see the "effort" to be carried out MS to try to put the nose on the ideal trajectory.

Believe it or not the car doesn't follow the path that set the wheels...

It is also important to clarify that this extreme situation occurs the first laps with cooler covers. Also this understeery behavior is repeated throughout the Grand Prix in a less noticeable way from the onboard camera, but it happens.

We see NR on two images of the same curve on the same lap. In the first, if we follow the direction of the wheels we would not end up in the grass?

But as shown below we continue on asphalt. And the angle of the steering wheel is still sufficient for the curve radius.

In the next picture we see the comparative chart of the data acquisition of the steering movements of MS and Johnny Herbert in their time at Benetton. The analysis is done on a corner of the circuit of Silverstone.

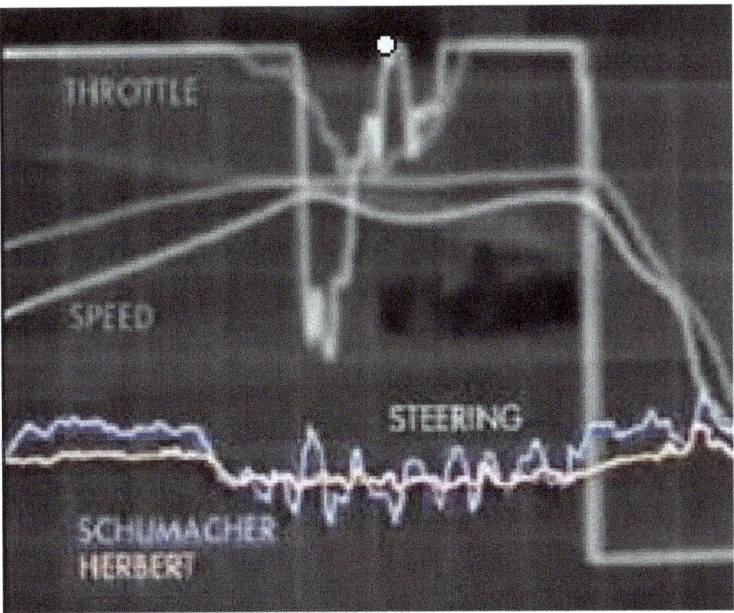

In blue Schumacher and red Herbert, we can see the most marked movements and the number of corrections made by MS against his partner at the time. This attempts to assert his aggressive driving style.

Surely there will be many things that could be discussed in this article, for example if this reason justifies the performance of MS in 2010, if the set up required to correct the understeer tendency makes car clearly inferior compared to the optimal (or at least respect of the his companion NR).

FULL EXPLOITATION OF THE SIDEPOD

In the 2014 season, Williams surprised by installing a system on top of the sidepod:

The reason for this slit-opening could be to attempt to exploit the entire front opening of the sidepod.

When air enters, due to the existence of the boundary layer and turbulence generated, the "useful" channel through sidepod is reduced:

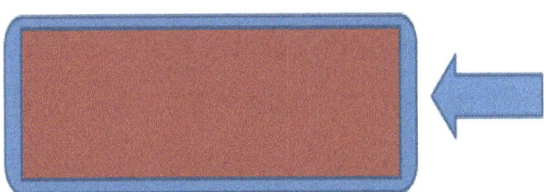

The slot, by means of the Venturi effect, sucks the flow moving through the inside, making the useful area expands; Moreover, it also increases the inlet flow with the same input area:

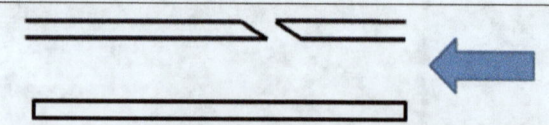

In case of no be able to design the car, the parameters of the car we can adjust are:

- Ride height
- Caster
- Camber
- Toe
- Springs
- Shock rebound
- Shock bump
- Shock piston
- Shock bump valves disposition
- Shock rebound valves disposition
- Roll center
- Antidive / antilift / antisquat
- Wings settings
- Tire pressure
- Antiroll bar
- Antiroll bar blades
- Antiroll bar blades position
- Bump rubbers
- Brakes proportion

$3^{(2\times19)} = 1\ 350\ 851\ 717\ 672\ 992\ 089$ posibilites.

If we add the possibilities or probabilities the driver produces due to his human nature as well as the different tracks and race conditions, the possibilities are virtually endless.

DRAG REDUCTION – CAR REAR

Have a rear vehicle; one system for reducing the drag, is to connect low pressure to zone that need not separate the flow; connect zone green and red.

The methods for that ? holes, cracks, canals, etc….:

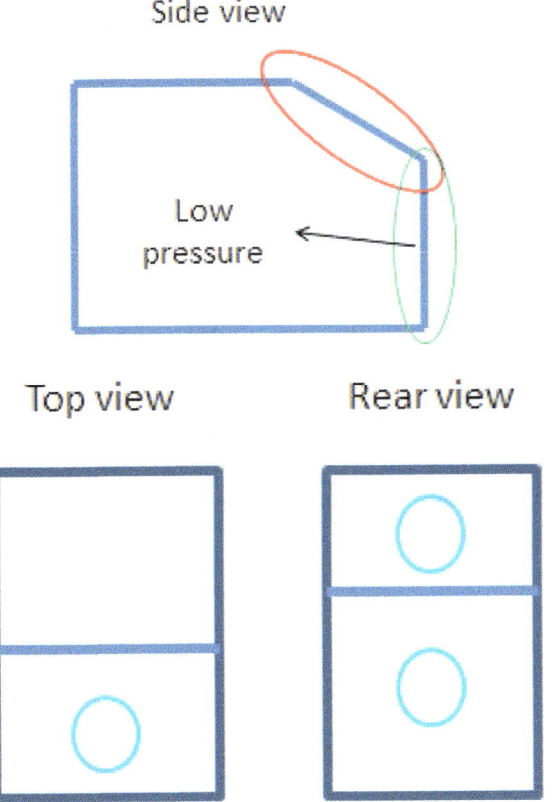

13.55. DRAG REDUCTION IN BOARDS ADVERTISIN GROAD

How to reduce drag in advertising boards that are placed on the road.

For this, the low pressure that occurs behind the sign is used; this depression sucks air directly hitting.

It is therefore possible to design a device side as follows:

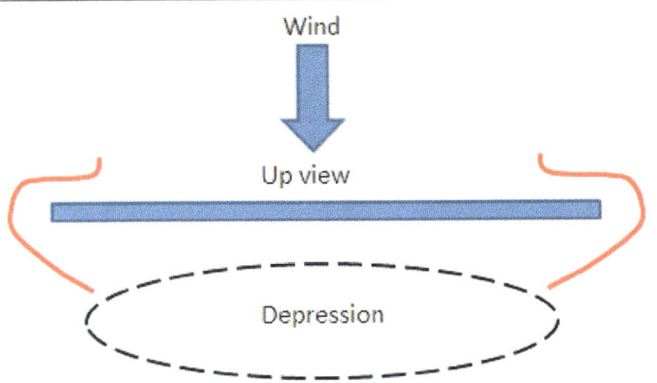

This device, have the same function than:

13.56. GEOMETRY MODIFICATION BY HEAT

Adrian Newey, for example, uses the heat of the exhaust gases, to modify the geometry or curvature of certain surfaces; the exhaust is properly focused:

13.57. IMPROVING DOWNFORCE IN CORNER

The principal goal of aerodynamicists in corner is augment the downforce in corner; that is complicate, yes, but is possible doing that:

When rear wing is active (if there is DRS – if not, is possible help the low pressure generate with the speed), help with the low pressure generate, to change direction air, by red duct; the blue duct, help to car to reduce the drag:

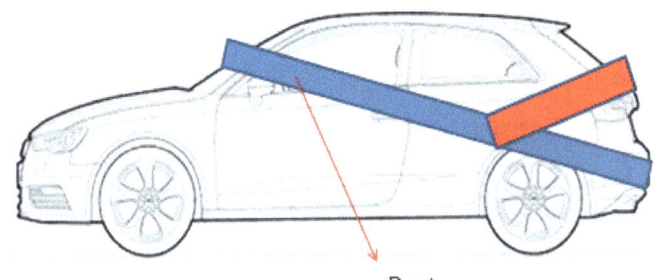

Duct.
Join high pressure front and low pressure rear.

When the air flowing through the red duct, is possible blowing this air in order to improve the downforce.

IMPROVING WING DOWNFORCE

This system, is have the same principle than Boundary layer blowing:

It allows the wing to achieve a higher incidence angle:

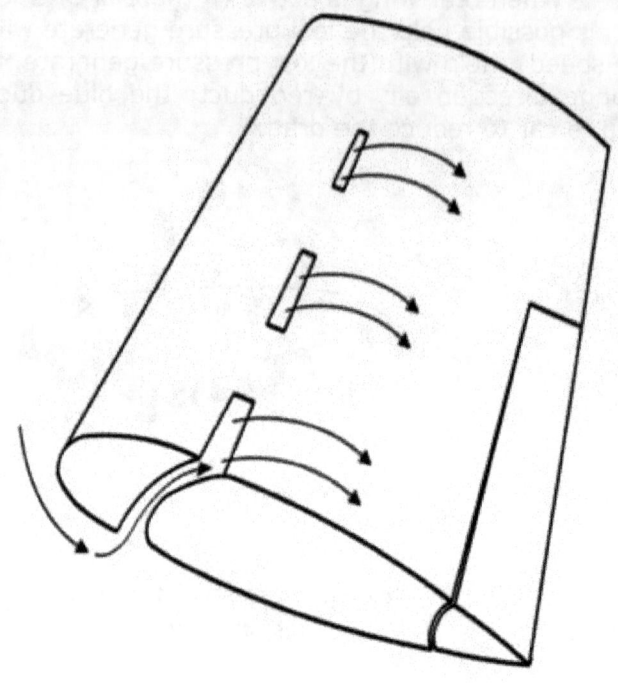

IMPROVING WING "EFFICIENCY"

The ideas are two:

- Reduce the drag: flow in blue duct.
- Augment the incidence angle: red duct; that produce one Boundary Layer Suction.

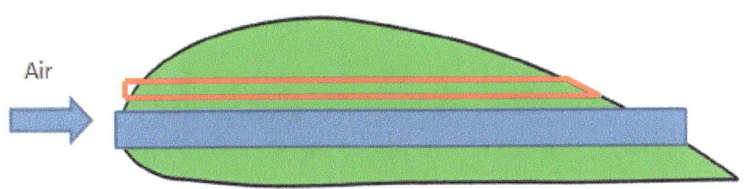

Is possible doing the same in normal car, reducing the drag:

13.60. REDUCING DRAG CAR

By the red duct, blowing air i i i i
That produces geometry "virtual" different (green):

. FERRARI 2016 SHAPE, AGAINST RED BULL TUNNEL

In 2016 season, Ferrari has a new rear design:

The idea is basically the same idea of Newey, about his Red Bull:

The goal: improve diffuser (also reduce drag and so, improve max speed).

OPTIMIZATION BEHAVIOR SUSPENSION - PORSCHE

Porsche developed years ago, based on the movement of the rear spoiler depending on the movements of the suspension system:

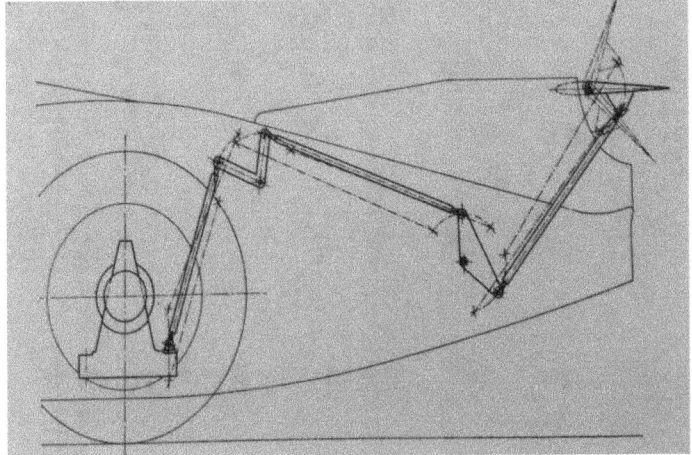

This system is very good because is possible through him, maintain the height over track; more: is possible, moving independent the two rear wings parts, maintain a good tire grip; this can be very important especially on oval circuits:

Another device about aerodynamic active:

S-DUCT – BOUNDARY LAYER .....

As we saw, the boundary layer produce drag (less speed produce friction and so, drag); so if is possibble to eliminate or reducing this boundary layer, the drag will be lower; we saw already the Boundary Layer Suction but also the Boundary Layer Blow:

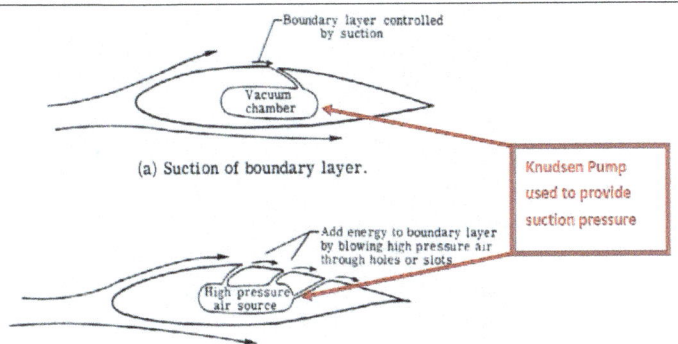

Vacuum chamber

(a) Suction of boundary layer.

Knudsen Pump used to provide suction pressure

Add energy to boundary layer by blowing high pressure air through holes or slots

High pressure air source

(b) Reenergizing the boundary layer.

Another system very important to reduce the boundary layer up nose cone, is the named "S-Duct":

Is possible to place the inlet below (as a Front Duct Ferrari) or in side lateral:

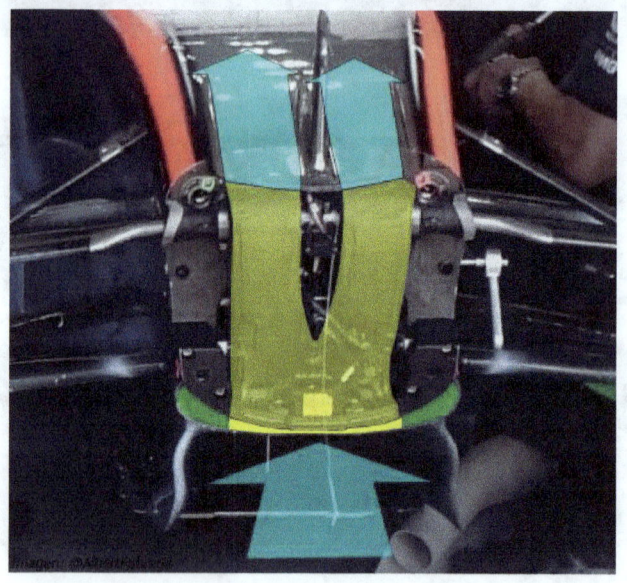

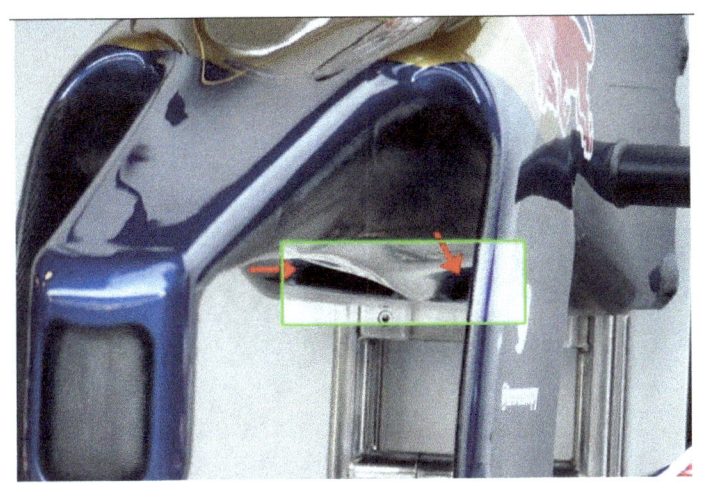

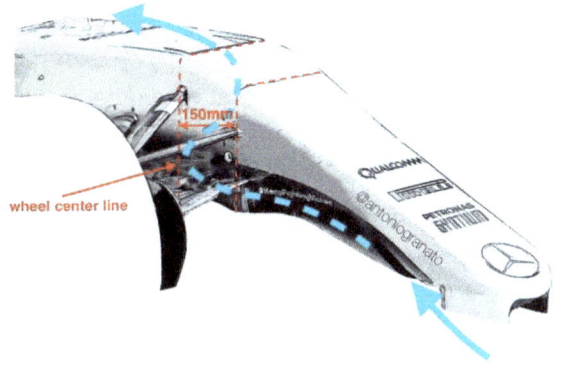

wheel center line

150mm

IMPROVING DIFFUSER

As we saw, is possible to place in diffuser edge, a gurney; that produce a depression witch suction the air below (optimizing diffuser):

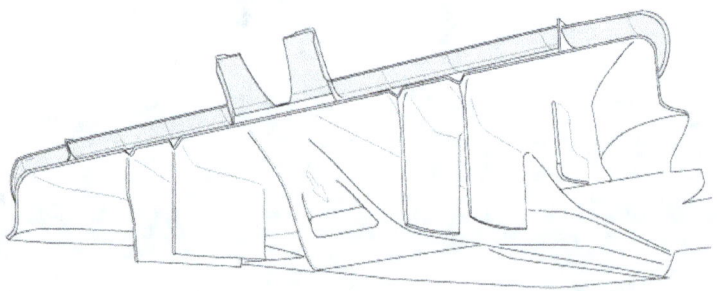

Is possible improve this system; amazing iiii:

Another device that is possible to place with the same main gal, is one wing active: that is: mobile:

GENERATION VORTEX: SKIRT - TUBE

We now already, what is the process for generating a vortex; that is: a low pressure tube.
If we are able to create a tube with holes, capable to suction air:

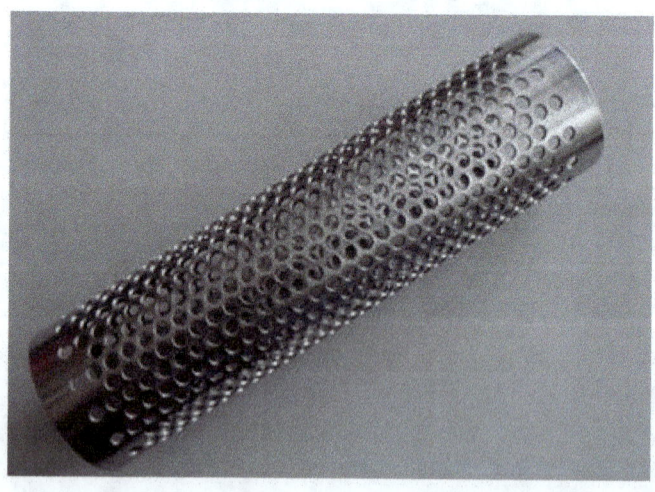

If this tube is placed in a car in movement:

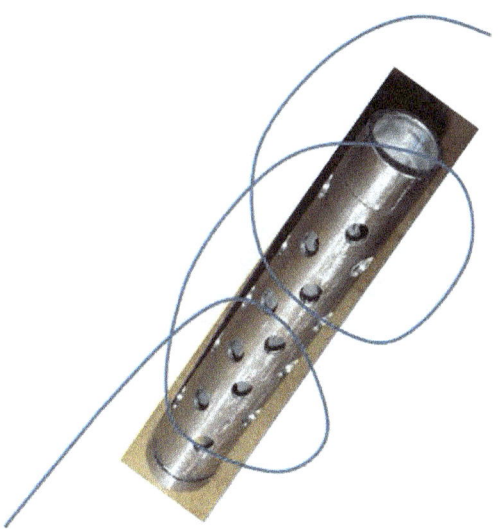

May be can be placed in ground, as a skirt "artificial"....

BARGE BOARD – IMPROVING IT

We know already the principal function of barge board:

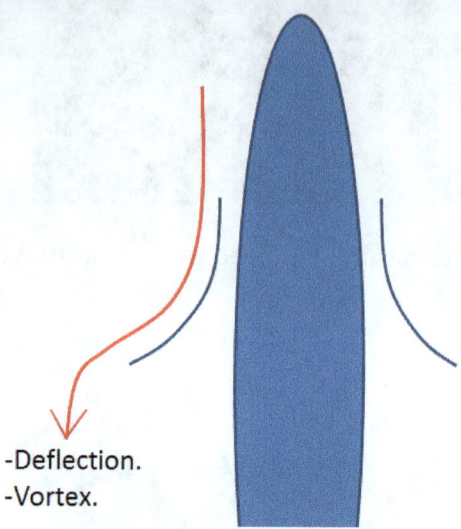

-Deflection.
-Vortex.

The problem is that produce a lot drag; if we remember some think about the ailerons, we can deflect air, with another way; that is:

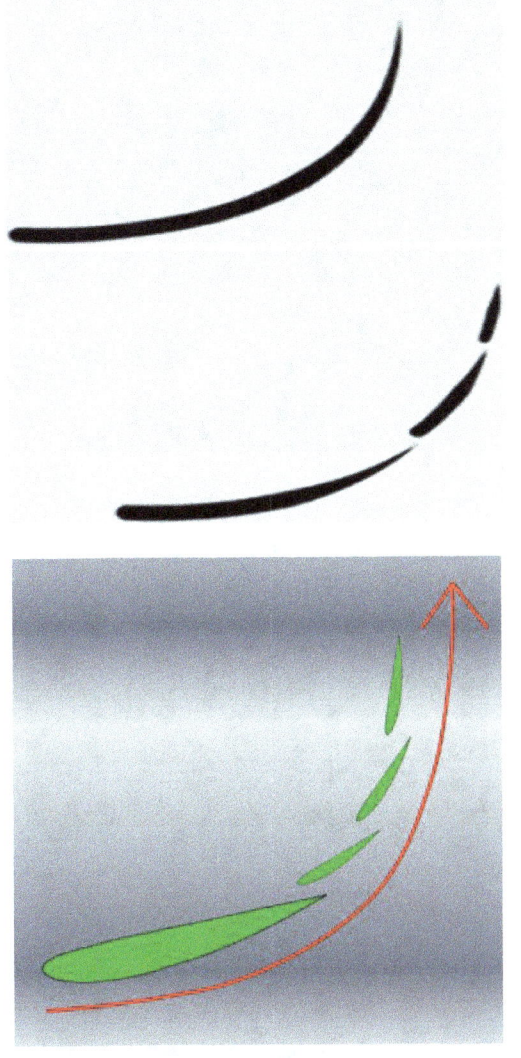

Williams in 2016 season, used that:

➔ Same effect, with low drag.

VENTURI FOR IMPROVING THE DIFFUSER

It will be possible to improve the diffuser, from Venturi canals installations up diffuser. That is:

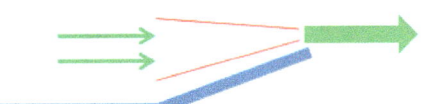

The air is accelerating and so, produces suction of air below diffuser. Produce drag, yes, and I think a lot drag, is possible, but the benefits are amazing.

FRONT WING – IMPROVING IT

We know already the principal function of front wig; but if have low drag, is better. For that, is possible to generate downforce with low drag, trough 4 or more flaps:

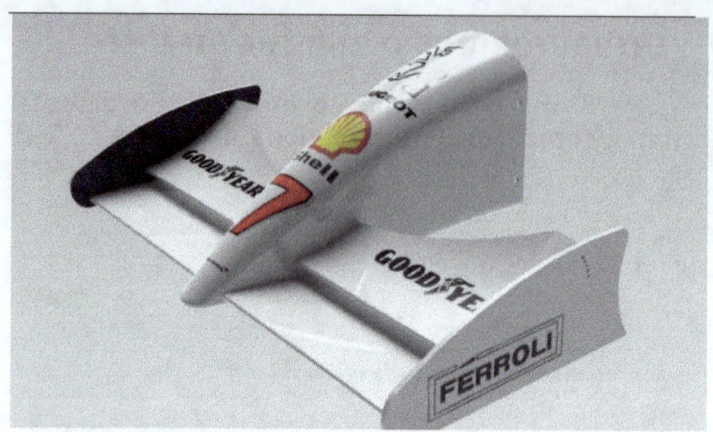

TIRE PRESSURE

The Mercedes - 2016 team has already found the solution to one of the small problems I had, which is the pressure of the tires in the race.

This system will explain and used it several pointers at the start of season equipment for the pressure of the tires get off the minimum in the regulations, which served to increase the productivity of the car when it is running.

The increased pressure in the tires was because of what happened in the Belgian Grand Prix last season where a back end, Sebastian Vettel busting his right rear tire on the climb to Eau Rouge, which caused a small battle between Ferrari and the Italian brand blaming each other who had erred.

In the grid at the European Grand Prix, it was shown the device, which seeks to increase the temperature in the brakes and tires to the 200c°, before mounting the corresponding compound. Once put the tire in the W07, it absorbs heat from the areas mentioned above, to achieve the minimum pressure that forces the regulation, all produced artificially; so that later, when the car is running, the temperature is reduced by half and, consequently, the pressure also decreases.

When the pressure is lower, it implies that there is greater surface contact with the asphalt admitting that the wheel cover can be folded. It also provides more grip and reinforces the tire to take heat from inside. However, when the pressure is high, the tread hardens and causes the tire is heated faster:

T-TRAY – BAT WING

There is a piece between car bod and ground or track, named Bat Wing.
Red Bull and Mercedes (2016), have different positions of this wing:

MERCEDES

In Mercedes is attached to body and Red Bull is attached to T-tray below. Why ????

Possibilities for Red Bull:

- In high speed, it does not allow more lower the T-tray, keeping the pitch.
- Allow to have springs softer.
- Play with the air flow….
- Etc….

Possibilities for Mercedes:

- Allow to have springs softer.
- Play with the air flow….
- Etc….

This piece produce two zones of pressure (low and high) that may be, to help Y250 Vortex…
Another interessant variation:

REAR WING "SLATS"

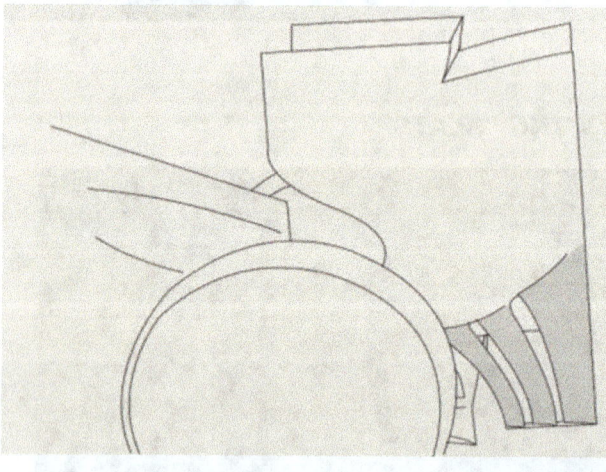

The concept was introduced by McLaren in 2011, starting with three large slats and an additional one mounted to the diffuser behind the rear wheel. This number has increased over the past year and the width of each has therefore decreased. Toro Rosso claim the biggest count of 11. The idea of having these here is to distribute the tyre squirt evenly out of the brake duct area and over the top of the diffuser.

I always try to imagine the purpose of this device by visualising a flow of water running down a pipe. If there is a small blockage in the pipe, the water will still flow over it but become turbulent. This also backs up the water behind it, slowing a larger section than anticipated around the blockage area.

This is what I presume is going on here. It takes the turbulent air ejected from the sidewalls of the tyre and straightens it out slightly, speeding the flow up behind the slats as a result and reducing the impact of the tyre squirt. The slats also help with the outwash of the rear wing, accelerating airflow at the base of the endplate to boost downforce. Outwash and upwash are ways of expanding the low pressure flow from beneath the wing to make it work better. This is why the endplates have a slight outward lip to them at their trailing edge.

REAR AND FRONT WING –SPOON- GEOMETRY

For those unfamiliar with the 'Spoon' design it can achieve more downforce from the central portion of the wing as it dips below the main planes intended legality window, whilst the outer sections being shallower help to retain a smaller drag footprint, meaning the car punches a different wake profile.

When the new regulations were introduced in 2014 they increased the previous Y75 zone to a Y100, in line with the exhaust placement regulations. This allowed the winglets (Monkey Seats) to be increased in size too, whilst some teams opted to run dual mounting pylons 100mm from the center line too, with the Mercedes W05 probably being the prime example, although Sauber have continued the trend this season.

What the regulations didn't do was adjust the centerline measurement for the main plane and top flap, even though it had been proposed, in order to mitigate some of the downforce that had been lost.   This changes in the new regulations though:

FIA RULE: 3.10.8 Any horizontal section between 600mm and 750mm above the reference plane, taken through bodywork located rearward of a point lying 50mm forward of the rear wheel center line and less than 100mm from the car center line, may contain no more than two closed symmetrical sections with a maximum total area of 5000mm2. The thickness of each section may not exceed 25mm when measured perpendicular to the car center line.

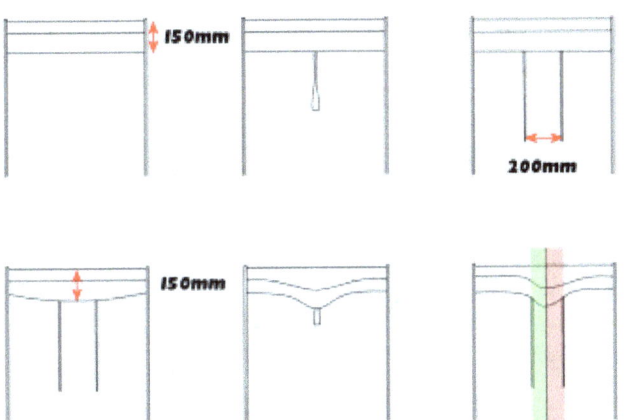

Upper-left is a standard wing using the full height of the legality box, no center support is used and so additional bracing would be needed at the wings base and/or thicker endplates which also carry the DRS hydraulics (Williams took this option in 2014).

Middle-Top is has the same legality box but uses a central pylon for stiffness and to carry DRS hydraulics.

Upper-right has the maximum legality box but uses twin pylons set at the maximum Y100 (200mm).

Bottom-right is a 'Spoon' style rear wing with twin mounting pylons, the maximum height is set at 150mm in the center but arches to a smaller height at the tips.

Middle-bottom exceeds the maximum 150mm in the center Y100 section of the wing and can have a range of flap sizes up to the maximum 150mm height at the tip.

Bottom-right shows the difference between the geometry that was available in green (Y75) vs the geometry that will be available in 2016, marking in red (Y100).

Haas team, have a little variation of this geometry; that is "DOUBLE WAVE"; the main goal is to reduce the drag, basically:

HIGH MASSFLOW IN REAR WING

The basic concept is adding mass flow air trough rear wing:
From front and from rear:

Is one possible option…. If the option is extract air from rear wing box, the goal may be, is extract air in order to improve the diffuser or another tsystem….

IMPROVING DOWN SIDE OF CAR WINGS

Is possible improve the flow in a down side in car wings; that is: flow with higher speed:

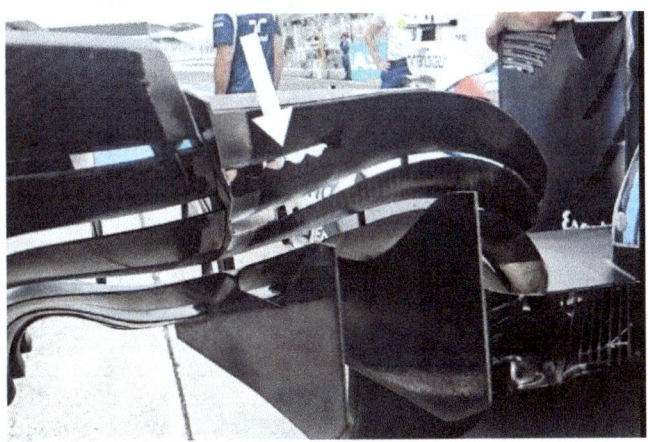

AIR INTAKE, AIR BOX AND ADMISSIONS

The main goal, as we know already, is that all outlets (trumpets) have the same air quantity; that is:

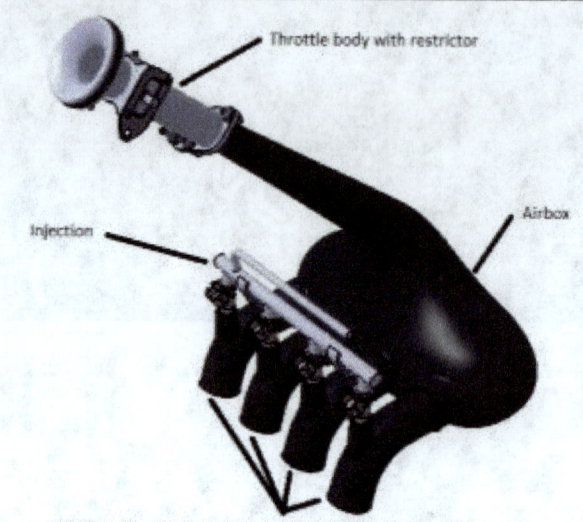

Throttle body with restrictor

Airbox

Injection

Outlets (Specified pressure, temperature, turbulence)

All this work is in CFD as a principal tool:

	Type	Polyhedral
Mesh	Number of cells	50 000 – 3 milion
	Number of prism layers	3 – 6
Solver	Type	Coupled Implicit, Flow, Energy
	Turbulence model	Realizable k-e, SST k-w, RSM
Physics	Wall functions	All y^+ wall treatment
	Max wall y^+	~ 50

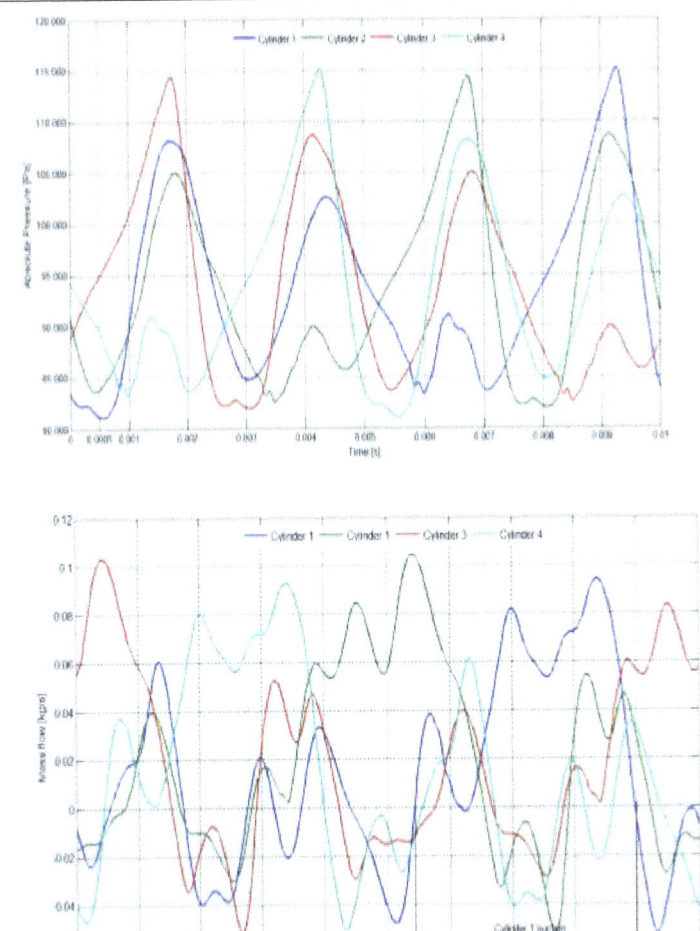

During cylinder 4:

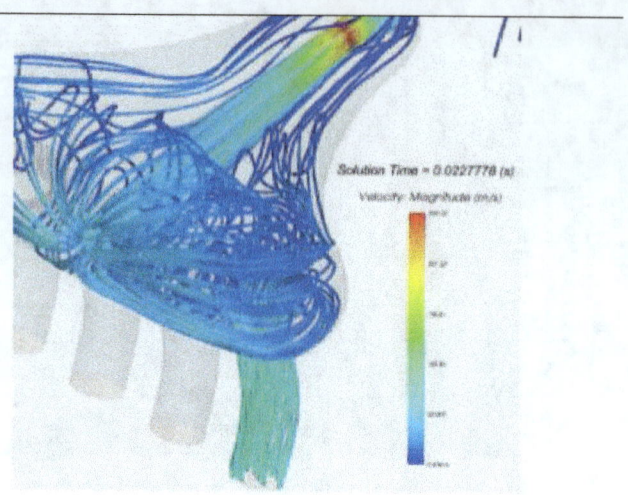

Solution Time = 0.0227778 (s)

Velocity Magnitude (m/s)

SUSPENSION ARMS IN X

The idea is simple: arms in X:

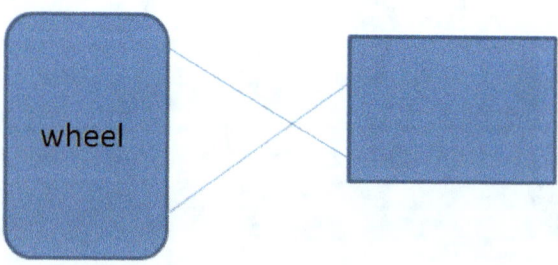

This geometry will be good in corner and in acceleration.

13.76. CRAKS FOR AIR ASPIRATION IN NOSE

The main goal is the air aspiration from front wing in nose:

That is the same than rear part in rear wing:

T-WING

Ferrari and Mercedes in 2017 season, introduce a T-wing in rear part:

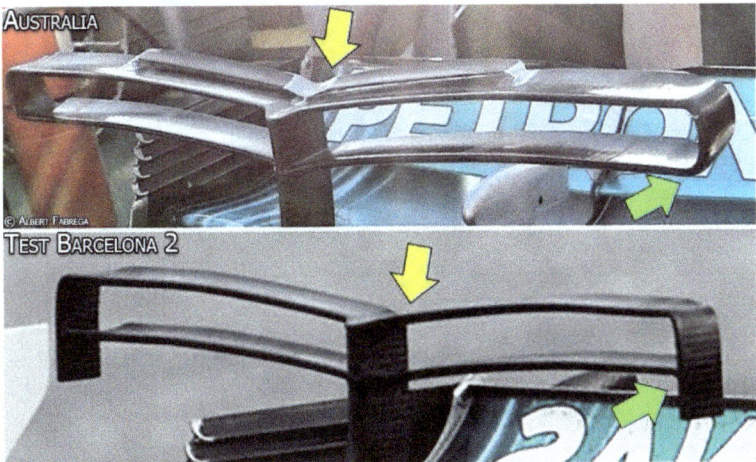

Looks a flow corrector to reduce the up deflection of the air leaving the Gurney flap of the R Wing to reduce Drag.

Generating downfore ??

FORCED AIR FLOW TO SIDEPOD

SHARK FIN FOR REFRIGERATION

This element, we know, have 4 goals:

- Understeering in corner.
- Reducing full drag.
- Improve downforce rear wing.
- Refrigeration:

In action:

SUSPENSION ARMS

Is possible to create the suspension arms, in order to:

- Reduce the drag or turbulences rear front wheels.
- Improve the refrigeration in sidepods.

For that, is built with the pickpoints more higher; that is:

**ARTICLES ABOUT AERO RACE CAR OR
VEHICLES ENGINEERING:**

Rake Angle on an F1 car

Enrique Scalabroni
Timoteo Briet
Nacho Suárez

1) Concept of Ground Effect in F1 cars

The bottom of the chassis or floor must be flat by technical regulations. It is the lower part of the chassis that extends from the beginning of the so-called tray, to the beginning of the diffuser. The lower surface of the 2 sidepods and even that of the floor itself, could originally be curved, as when Colin Chapman created the Lotus 78 and Lotus 79 (Figure 1).

Figure 1 – Curved floors under the sidepods of an F1 car

When a car travels at a certain speed on a fixed surface (asphalt surface of the circuit) in a gaseous medium such as air, and there is a certain clearance distance between both surfaces, there is a significant reduction in static pressure (below the atmospheric pressure) between the duct walls and the circuit surface, which generates a vertical force or "**downforce**" (DWF) that pushes the chassis down. This force is proportional to the static pressure on the surface of the bottom of the car and its two sidepods, multiplied by the total section of that plate. This effect, due to the speed of the car that generates a negative pressure in the conduit with respect to atmospheric pressure, is known as the "***Ground Effect***".

We must bear in mind that the reason for the Ground Effect occurs for three physical reasons or numerical simplifications of reality: the Bernoulli's Principle, the Continuity Equation and the Venturi Effect (which derives from the two previous ones).

The diffuser starts where the flat floor ends, increasing the section or passage area. The diffuser is in charge of "*diffusing*" the pressure and increasing it gradually until it reaches the atmospheric pressure value that surrounds the car.

The continuity equation describes the uniformity in terms of flow within a duct or tube. If there is no opening within that conduit that allows fluid to enter or exit the conduit. The mass-flow rate of air entering the conduit will be equal to the mass-flow rate of air leaving the conduit. Recall that we are dealing with speeds of the car that are of the order of Mach 0.3, so the fluid can be considered as incompressible.

Considering the Bernoulli's Principle, for a constant mass-flow rate, the total energy remains constant along a flow line (or "*streamline*") through the duct, so if the cross-section area decreases in one part of the duct, the air speed will increase in that part and, therefore, the pressure will decrease.

The downforce generated by the ground effect is very important because it is a force that does not have inertia. Furthermore, this force is essential since it compresses the elastic part of the suspension system and consequently compresses the tyres against the ground, increasing the friction force between the contact patch and the asphalt.

It should be remembered that before the ground effect was used, F1 cars generated accelerations of approximately 2 g while cornering and 2.8 g while braking, whereas the Lotus 79 was very effective and generated 2.8 to 3 g and 3.5 g respectively. Those values created a great difference in efficiency and from that moment, that radically changed the aerodynamic design of racing cars.

It is important to indicate that it is **NOT** necessary to use a curved floor, because the most effective system is to use a flat floor, followed by a diffuser, because it allows a greater stability of the air flow that circulates under the car and maintains a lower average negative pressure. This was the reason why the Williams FW07 were more efficient than the Lotus 80, as Patrick Head was the first Engineer who designed the flat floor on both sides and a long diffuser with an angle of maximum 7 to 10 degrees, to keep the boundary layer adhered to the diffuser walls.

The pitch or "*rake*" is defined as the angle that the car's floor forms with the asphalt; the greater this angle, the greater the downforce the car will generate. But this angle has a limit for each car design. In order to achieve a high rake angle, it is necessary that both the floor and the diffuser work together optimally and this, logically, is complicated. There are many variables that intervene in this optimization: dimensions of the floor and diffuser, inlet and outlet areas of the floor and diffuser, heights, expansions ratio of the floor and diffuser, etc.

A bad design in the floor-diffuser assembly can produce the so-called "*porpoise effect*". It occurs when a viscous air blockage under the floor is created at high speed, impeding more air from passing through the floor and, hence, losing downforce which, in turn, increases the clearance height and then it generates a high vertical force again. This creates a sinusoidal movement with a certain frequency that creates stability problems in the car, a problem that forced Colin Chapman to use very stiff springs to control this oscillation at high speed and in fast corners, which in turn created the additional problem of making the tyres work with high oscillation amplitude, not allowing to solve the problem.

As a summary, the air circulation below the bottom of the car, follows the following path: it enters from the front of the floor and accelerates as the cross-section area is reduced; it continues under the floor producing low pressure which sucks the car down (*downforce*); it reaches the diffuser and exits the rear of the car. But how is it possible for the diffuser to extract or produce force to pull the air flow if there is "more pressure" in the diffuser than under the floor? The answer is that, at the transition between the floor and the diffuser, right at that kind of corner, the so-called "*crack pressure*" is occurs, which, as its name indicates, generates a very sudden pressure drop, which is the responsible for pulling the air flow from the front of the floor, forcing the air to pass through the entire floor.

In Figure 2 and 3 there are some views of the car (a Formula-E) used to verify the pressure distribution under the floor-diffuser system explained above, by means of CFD analysis (test conditions: air speed = 250 km/h, rake angle = $1°$, rotating tyres and rims).

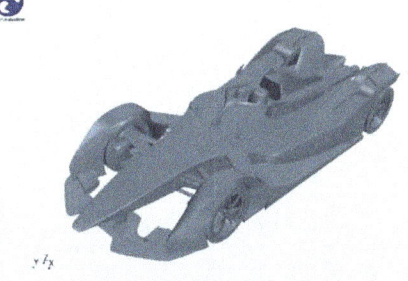

Figure 2 - Formula-E type car CFD model

Figure 3 - Flat floor and diffuser of a Formula-E car CFD model

Figure 4 shows the pressure distribution at the centre line under of the car, where "A" corresponds to the floor section, "B" is just the aforementioned "*Crack Pressure*" or "*braking line*" and "C" is the diffuser section. It can be notice exactly what has been explained before: "B", a noticeable low pressure, a greater average pressure in "C" than in "A"; in fact, the diffuser produces little total load, compared to the floor.

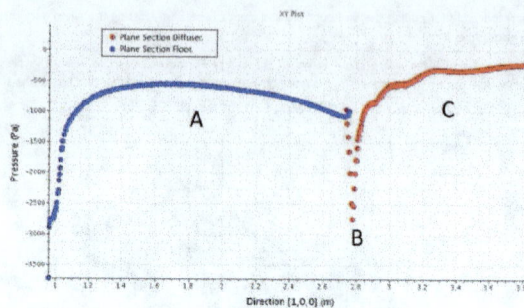

Figure 4 – Pressure distribution at the floor-diffuser assembly of a Formula-E car CFD model

A surface (static) pressure plot is shown in the bottom view of the Formula-E model depicted in Figure 5 where the aforementioned zones (A, B and C in Figure 4) can be clearly distinguished. The low-pressure areas are represented in coloured blue, that is, the beginning of the floor (where the air flow accelerates) and the Crack Pressure area at the separation line between the floor and the diffuser.

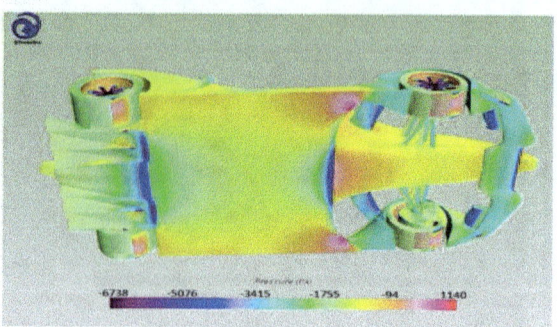

Figure 5 – Surface (static) pressure plot (bottom view of a Formula-E car CFD model)

Another surface (static) pressure plot, with a different car view, can be seen in Figure 6.

Figure 6 – Surface (static) pressure plot (top view of a Formula-E car CFD model)

2) Global analytical comparison between a *"double-setup"* car (high rake), compared to a *"single-setup"* car (low rake)

The analysis is carried out based on the setups, represented in Figures 7 and 8 (1° rake angle) and in Figures 8 and 9 (3° rake angle), that are imposed on a basic model of an F1 car.

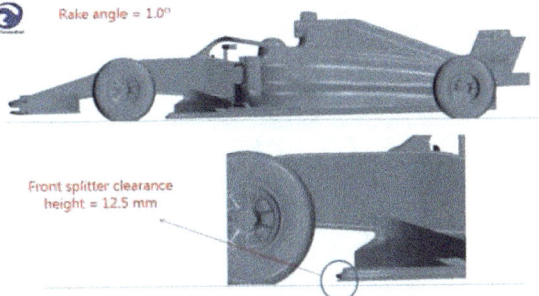

Rake angle = 1.0°

Front splitter clearance
height = 12.5 mm

Figure 7 – Simple Formula 1 car CFD model (1° rake angle setup, lateral view)

Rake angle = 1.0°
(frontal view)

Figure 8 – Simple Formula 1 car CFD model (1° rake angle setup, front view)

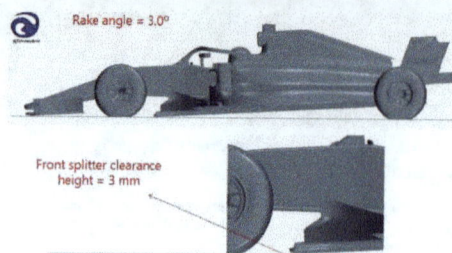

Rake angle = 3.0°

Front splitter clearance
height = 3 mm

Figure 9 – Simple Formula 1 car CFD model (3° rake angle setup, lateral view)

Rake angle = 3.0°
(frontal view)

Figure 10 – Simple Formula 1 car CFD model (3° rake angle setup, front view)

Other images of the analysed car are shown in Figures 11 through 15. Technical data of the CFD simulations is exposed at the end of the document.

Figure 11 – 3D view of a simple Formula 1 car CFD model (front view)

Figure 12 – 3D view of a simple Formula 1 car CFD model (rear view)

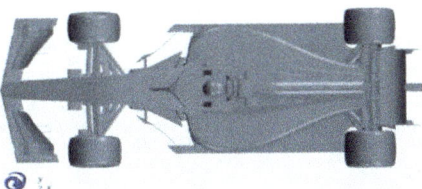

Figure 13 – 3D view of a simple Formula 1 car CFD model (top view)

Figure 14 – 3D view of a simple Formula 1 car CFD model (bottom view)

a) Car in high-rake condition (*double setup*)

The system of a high rake or aerodynamic double-setup racing car consists of a car that presents a basic change in the **Concept** of the racing car design, since it is a car that works with two aerodynamic setups in the same car, so it needs a rear suspension system that works with a double vertical stiffness and a front suspension that allows a precise control of the clearance height, provided with a combined damping system to act in the 3 states: 1) movement in heave (vertical movement of the suspended mass without rotation in the Y axis) and in braking (controlled by the 3rd element); 2) movement in a curve, to damp the movement in roll; and 3) movement to control the damping of each wheel at high frequency and the impact against the kerbs.

The elastic system of the front suspension is provided with two torsion bars installed on the shaft of each rocker, which support the static load of the car and the variations in dynamic loads from the downforce, the circuit disturbances and the braking, cornering, acceleration states and their combinations, allowing a precise control of the clearance height. The group of shock absorbers allow the accumulated energy to be dissipated by the different load variations that the car undergoes in its movement along the circuit. It also has an anti-roll bar to minimize the roll angle of the car.

It should be noted that all these elements are commonly used for several years in competition cars, but, in the case of this car, they must allow a correct operation both in the low-rake state (in the straights) and in the high-rake state (at the initial braking moment and when cornering, specifically in low-speed corners), because this car allows a large variation of the rear clearance height.

The rear suspension also has similar elements to control the height of the rear axle, firstly the control of the static load, plus the variation of the DWF and the variations of the dynamic loads when moving the vehicle on the circuit, similar to the case of the front axle, but with different values due to the distribution of static and aerodynamic loads, plus the control of the clearance height of the rear axle, due to the considerable variation in height when driving the car on straights and when braking and cornering (which is a function of the car speed). The car must have a suspension with two different values of vertical stiffness, a soft one, to work at a clearance height that ranges from 165 to 85 mm of the rear sprung mass and another with greater vertical stiffness to control the height of the rear sprung mass when it is between 85 and 45 mm.

When braking, before entering low-speed corners, the vehicle increases the rear height due to both the load transfer and the loss of speed. This reduces the DWF, increasing, at this time, the height of the rear Centre of Gravity, reaching approximately 165 mm rear height and 3 mm of front splitter clearance. Due to the longitudinal weight transfer under braking, the lower front lip of the chassis floor approaches the pavement surface, to the point of hitting or leaving a 3mm gap, reducing the clearance height of the front wing, increasing the front DWF, due to the Ground Effect. That reduces the possibility of understeer in the front axle by means of the 3rd element of the front suspension that does not allow the clearance height of the front wing to touch the asphalt. The roll moment created by the centrifugal force, acting on the sprung mass at the height of the Centre of Gravity, which increases due to the greater clearance height of the rear axle, must also be controlled (it should be remembered that by increasing the rake angle, the height of the engine and the gearbox increases, relative to the asphalt surface due to the roughly 3-degree rake angle that the high-rake car takes).

When driving the car in low-speed corners, it generates a high lateral weight transfer from the inner to the outer wheel one on the rear axle, due to the high roll moment (as mentioned above), unloading the inner wheel and causing the car to turn practically on three wheels (the two front wheels and the external rear wheel). That facilitates the car to turn, due to the reduction of the mechanical grip on the rear axle, because it practically works on the external wheel (which has a grip potential of the order of 60 % to 70 %, according to the radius of the curve). To avoid slipping of the internal wheel (*wheelspin*), it is necessary to use a high percentage of locking in the self-locking differential (approx. 60 % to 65 %). To prevent the front wing, which is very close to the ground, from touching the ground (as this would cause it to lose downforce) its height is controlled by the third element and its roll is controlled by a rather stiff roll bar (in order not to lose downforce on the part of the wing on the inside of the corner).

Ultimately, the concept of high rake allows the reduction of understeer in low-speed corners. With this, the car has a controlled lateral grip (at the limit of lateral slipping of the rear train), controlled by braking deceleration and the steering wheel angle given by the driver, thus keeping the percentage of oversteer (*controlled oversteer*) of the rear end and minimizing the understeer (*min. understeer*) of the front end. So far, not all drivers can drive this type of car as correctly as Max Verstappen can, because Adrian Newey allowed him (almost from his start in F1) to learn and develop the driving technique of a high-rake car (Figure 15).

Figure 15 – Adrian Newey (left) and Max Verstappen (right)

For this reason, the high-rake car can be very fast in low-speed corners, by allowing a fast change of direction, without practically understeer. But there is the problem that, as the cars works with practically a single wheel on the rear end in low-speed corners, the external rear tyre is overstressed, which increases external tyre wear, probably causing a greater degradation than in a traditional low-rake car. The latter has a smaller vertical load difference on both rear tyres in low-speed corners, and does not overstress the rear tyres when cornering, allowing a greater number of laps with the same tyre set compared to a high-rake car. Low-rake cars put more stress on the front axle tyres, because they have a greater clearance height in the front wing in low-speed corners and this causes less ground effect than the high-rake cars do, so they are closer to the critical understeer.

Analysing now the high-rake car during the moment of acceleration, at the exit of low-speed corners and at the start of a race together with its behaviour in a straight line at high speed, the following points are observed:

- In the case of starting the race (on dry or wet ground), as well as at the exit of low-speed corners, the high-rake car has the clearance height in the rear end of around 165 mm at the start of the race and between 145 and 150 mm at the exit of low-speed corners. So it is understood that the rear end is still supported by the elastic system with less vertical stiffness (remember that the high-rake car has two vertical stiffness values as explained above). This makes it possible to generate a greater linear acceleration, because as the rear end is supported, at that moments, on a low-stiffness suspension system, at no time will the car be at a point of high vertical stiffness in the rear axle, which would cause the rear tyres to skid, losing longitudinal acceleration, since the softer suspension with allows higher longitudinal acceleration, with greater progressiveness in the application of power under acceleration. This allows a more efficient acceleration on wet floors and floors with little mechanical grip.

- At the moment of acceleration (when changing to higher gears), it also allows a greater acceleration efficiency between gears. But remembering that the DWF grows in quadratic proportion to the increase in vehicle speed, the increase in DWF will quickly compress the softer elastic system of the rear suspension, leaving the car, after the 5th gear, supported by the stiffer elastic system, so the rake angle gradient will begin to tend at a constant angle of approx. 1.0 to 1.3 degrees. As the DWF continues to increase, that will bring the clearance height of the rear sprung mass below 70 mm, reducing the "$C_d \times A$" product and the angle of incidence of the rear wing, which reduces the drag of the car as a result of the reduction of the car's master section (A) and its C_d coefficient. At the same time, the DWF on the rear axle increases due to the increase in ground effect because of the proximity of the asphalt tape to the rear diffuser due to the reduction of the clearance height of the rear axle.

- The reduction of the "$C_d \times A$" product increases the vehicle speed and the car enters a similar aerodynamic setup field of a traditional low-rake car, so the car does not recover the consumed power compared to low-rake cars, where the final speed depends on the BHP value of the power unit (PU). Therefore, if a traditional car has a higher BHP value, this car will be faster on straights, because, as both cars have a similar "$C_d \times A$" product. The highest speed will be performed by the car with the highest power, although the high-rake car is faster in slow corner zones.

- In high-speed corners, there is not big efficiency difference between both cars (traditional or low-rake cars and high-rake cars), because both will have a very similar rear axle clearance height. In the high-rake car, its rear sprung mass will be already supported by the stiffer elastic suspension system, which is similar to that of the traditional car and the DWF is high and very similar. Therefore, the rear diffuser will generate a very similar ground effect value, as well as the rear wing, because the angle of incidence at that position of the circuit is quite similar (taking into account that, in order to simplify the study, we are considering that both cars work with similar ailerons that generate similar DWF, drag and aerodynamic efficiency).

b) Car in low-rake condition

To simplify the study of the double-setup car in comparison with a single-setup car (those with low-rake aerodynamics), an aerodynamic study in CFD has been carried out, considering the same car and varying only the rake angle from 1.0 to 3.0 degrees to determine the aerodynamic response to the change in rake angle, as it happens in high-rake cars in braking moments and in low-speed corners. On the other hand, a change in the rake angle from 3.0 to 1.0 degrees has been considered for the car when driving on high-speed straights (where it works with a low rake angle). Only CFD simulations in both angle values has been performed, that is, it has been considered two fixed and constant states.

It is important to note that, when varying the rake angle from 1.0 to 3.0 degrees, the aerodynamic *"Centre of Pressure (CoP)"* is shifted forward (in this particular case, from 43.4 % at the front for the 1.0° rake angle to 53.1 % at the front for the 3.0° rake angle). As can be observed, by varying the ground effect value in the front wing (and in the front part of the floor) due to the reduction of the front clearance height and the consequent loss of the DWF in the rear axle because the increase in clearance height in the rear end, makes the rear diffuser lose ground effect, so the aerodynamic balance varies, due to the proportional increase in the DWF in the front end and the loss in the rear end.

In Figure 16 surface pressure plots are depicted for a F1 car CFD model with a rake angle of 1.0° (top view) and 3.0° (bottom view). Low-pressure areas are represented in coloured blue and high-pressure in red. It is evident that at 3.0 degrees, blue areas increase substantially, indicating that the front wing is generating more downforce due to ground effect.

In addition, not only does the load on the front wing increase, but as said before, so does the lower part of the *nosebox*, that works as a front diffuser. With this, the CoP position with respect to the X axis can be calculated, where it is observed that, in the 3.0° rake angle car, the CoP moves forward with respect to the CoP position of the 1.0° rake angle car (Figure 17).

This difference in load distribution, because of different CoP location, can be appreciated by calculating the streamlines underneath the floor-diffuser assembly (Figure 18).

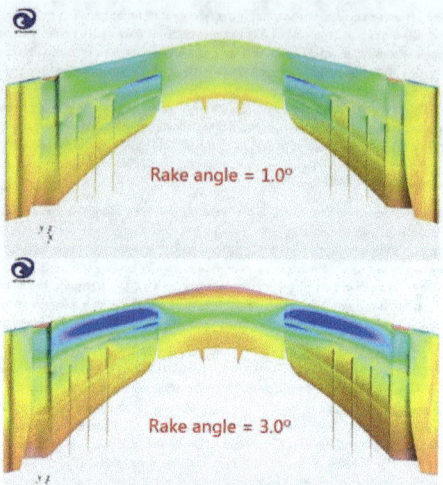

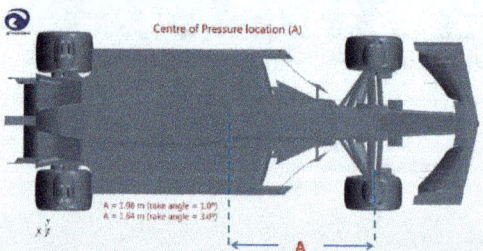

Figure 16 – Surface (static) pressure plots at the bottom of the front wing of a F1 car CFD model at 1.0° rake angle (top) and 3.0° rake angle (bottom)

Figure 17 – Centre of Pressure (CoP) location on the F1 car CFD model

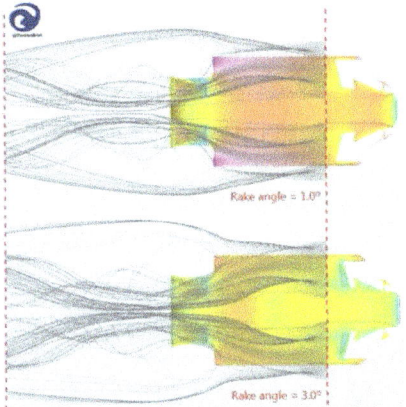

Rake angle = 1.0°

Rake angle = 3.0°

Figure 18 – Streamlines comparison for the F1 car CFD model

It can be noted that the double-setup car (the one with 3.0° rake angle), when driving on straights, or when exiting low-speed corners and increases speed, or also in fast corners, reduces its rake angle, but in these corners has a slightly higher rake angle than the single-setup car. Analysing in this way, it possible to clarify the physical fact of the aerodynamic behaviour of the car when varies the rake angle because of the increase in DWF as a consequence of the increase in vehicle speed. Looking at the pressures underneath the bottom of the car, in particular at the diffuser, large differences are observed. Three different pressure distributions have been calculated along three longitudinal sections (A, B and C) as shown in Figure 19.

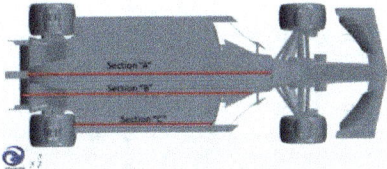

Figure 19 – Sections used for computing pressure distributions in the F1 car CFD model

The resulting pressure distributions are shown in Figure 20 (for the case of the 1.0° rake angle car) and Figure 21 (for the 3.0° rake angle car). Starting from the left of the plots, the first part of each curves corresponds to the beginning of the flat floor, whereas the end part corresponds to the end of the diffuser (if applicable).

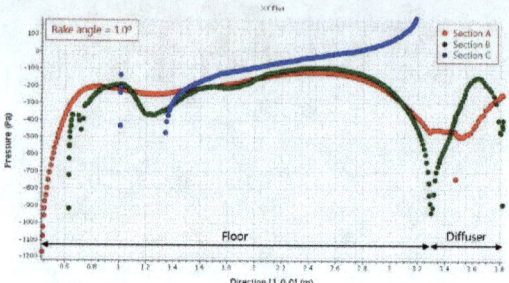

Figure 20 – Pressure distribution at different sections of an F1 car CFD model (rake angle = 1°)

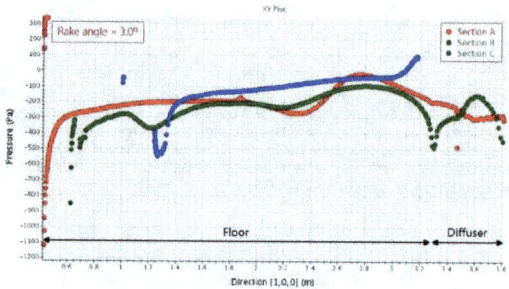

Figure 21 – Pressure distribution at different sections of an F1 car CFD model (rake angle = 3°)

One of the most interesting things that can be seen is the great low crack pressure (green peak down at the transition between the floor and the diffuser) that occurs in the case of 1.0° rake angle (Figure 20). On the contrary, in the case of 3.0° rake angle (Figure 21), the downward peak is much smaller and, hence, its crack pressure value is higher (less negative). As explained before, in the latter case, the diffuser does not work as well as in the former case, due to the greater clearance height at the rear axle of 165 mm due to the rake of this setup. In reality, the floor-diffuser assembly generates less DWF than in the case of the 1° rake angle car. This difference in pressure distribution is also reflected in a different distribution or evolution of the streamlines underneath the car floor as shown in bottom (Figure 22) frontal (Figures 23 and 24) and rear views (Figures 25 and 26).

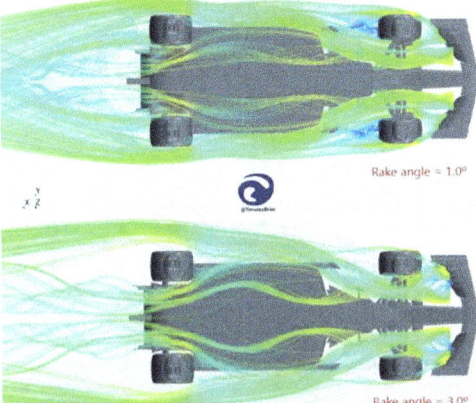

Figure 22 – Streamlines comparison for the F1 car CFD model (bottom view)

Figure 23 – Streamlines comparison for the F1 car CFD model (front view, rake angle = 1°)

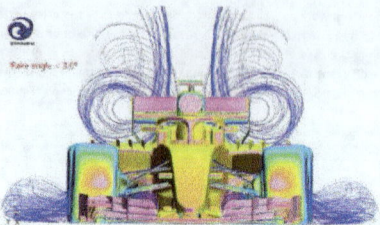

Figure 24 – Streamlines comparison for the F1 car CFD model (front view, rake angle = 3°)

For that reason a single car model is used as a reference model, without making modifications or aerodynamic corrections in parts of the body, ailerons and/or flaps, nor are special developments carried out to improve the efficiency of each set-up (whether with high or low-rake) only the response of the aerodynamic behaviour of the car is evaluated in its values of total DWF, drag, aerodynamic efficiency, percentage aero-balance of the front and rear DWF (with respect to the values of the clearance heights of the front and rear axles), the pressure diagrams on the car body and pressure fields around the car, diagrams of velocity, vorticity, position of the CoP (for high and low-rake conditions), streamlines at the top and the bottom of the car and also on the surface of the circuit track, vortex generations and other additional verifications, to determine the maximum efficiency of each set-up.

It is observed that this type of double-setup cars behave like a total DRS (Drag Reduction System), different from the FIA DRS that acts only on the rear wing to reduce the angle of incidence of the upper flap and reduce the drag of the rear wing, increasing vehicle speed to allow it to overtake the car in front. The angle of incidence of the double set-up car, varies mostly with the increase of the vehicle speed, with respect to the single setup cars, that have a much smaller variation of the angle of rake.

Remembering that the high-rake car (when it is braking or when cornering in low-speed corners), thanks to the reduction of the front wing clearance height with respect to the track surface, is able to generate a higher DWF value in the front axle, due to the increased ground effect on the lower part of the front aileron, its flaps and the lower part of the nosebox. Because of its inclination, the entire front wing assembly works as a front diffuser, rather than a wing, allowing to brake more deeply into the corner and to reduce the degree of understeer. This reduces the braking and traverse times on low-speed corner, due to the increased driving speed with the car under control. In this way, it is possible to anticipate the application of power on corner exit, as the higher DWF of the front wing means that it does not suffer from critical understeer on entry and exit of slow corners, allowing it to accelerate before fully exiting the corner. At this point, the car has a lower vertical stiffness in the rear axle than the stiffness on a straight line (and a lower value than that of the low-rake cars), which allows the vehicle to develop power in a more progressive way, without reaching the sliding limit of the rear tyres (as explained in the first part of this report).

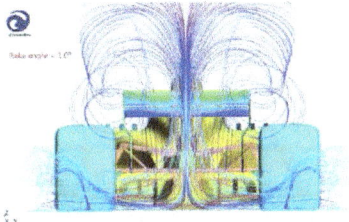

Figure 25 – Streamlines comparison for the F1 car CFD model (rear view, rake angle = 1°)

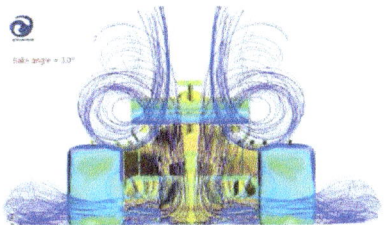

Figure 26 – Streamlines comparison for the F1 car CFD model (rear view, rake angle = 3°)

In other words, the potential advantage is at the moment of braking, where the high-rake car, in the first moment of braking, increases the rake angle almost instantaneously, thereby increasing the drag by increasing the rear clearance height, increasing the incidence of the chassis and especially that of the rear wing. The car is very effective when braking and reduces the possibility of locking the rear tyres because the car is, at that moment, supported by the softer of the 2 vertical stiffnesses of the rear axle.

c) Final considerations

This is just a basic study to understand the dynamic effect of a double-setup (high-rake) car versus a single-setup (traditional low-rake) car. There is still a lot to study, analyse and develop in order to achieve a highly efficient double-setup car. Considering that Adrian Newey began to develop this idea in 2009 and has already been working on this concept car for almost 12 years, it is very difficult for other designers to achieve a similar efficiency in their projects, since a lot of study and development work is needed to define an excellent vehicle dynamics: in aerodynamics, in the study of lateral load transfer in low-speed corners and when braking and accelerating, to make this double-setup car system work properly.

Also, as a final consideration, the numerical results of both simulations, and their variations with respect to the change of setup (rake angle), are presented. Many conclusions can be drawn from this first and quick analysis. Especially it can be seen that varying the rake angle from $1.0°$ to $3.0°$ the front DWF increases, however, the total DWF suffers a small variation (of the order of 1.7 % lower), so the efficiency in the car with $1.0°$ rake angle (since the drag is much higher than that of the $3.0°$ case) is bigger than that of the car with a $3.0°$ rake angle (see Table 1). It is also observed that the increase of front DWF shifts the CoP forward.

Table 1 – Numerical results of simulations with a F1 car CFD model at two rake angles

Comparative Analysis	Rake Angle	
	1°	3°
Front Wing Downforce		+ 39.28 %
Floor-Diffuser Downforce		- 4.1 %
Total Downforce		- 1.7 %
Aerobalance (front)	43.4 %	53.1 %
Total Drag		+ 15.82 %
Efficiency	2.5	2.1

Note that the car on which the analysis has been performed is not optimised to work with a $3.0°$ rake angle; therefore, by carrying out a good development and optimisation work, the total DWF and the aerodynamic efficiency could be increased considerably.

With regard to the efficiency of the Red Bull car, it is observed that, in order to be as efficient in straight line speed as the Mercedes F1 car, it needs an increase of the current power output of its engine/PU by an estimated value of 4 to 4.5 %. It remains to be seen whether this will be achievable by Red Bull's engine/PU supplier during the current championship.

It is interesting to note that the teams that have developed high-rake cars have lost less lap time compared to last season. The teams that have kept the low-rake concept, by changing the FIA 2021 regulations (basically by cutting the floor size), have lost even more lap time. Whilst Red Bull has lost around 1 second of lap time, Mercedes has lost around 2 seconds or so.

Being able to increase the rake angle of the car, allows to increase the DWF, because, if the FIA regulation reduces the DWF load in the order of 10 %, those teams that designed cars "capable" of having high-rake angles, can compensate for the loss of load by increasing the rake. Simple, isn't it?

Data of each CFD simulation

- Test speed: 250 km/h.
- 65 million cells.
- Boundary layer composed of 20 layers.

- Rotating tyres and rims.
- Moving floor.
- Radiators on sidepods.
- Engine intake.
- Gas mixture and temperature in the exhausts: 120 m / s.
- Front wheel brake disc and calliper.
- Heat transfer in engine block, exhausts, radiator and brakes.
- 72 hours of calculation on a PC with 256 GB RAM and 56 cores.

About the authors

Enrique Scalabroni

Dallara Automobilli, Ferrari F1 Chassis Technical Director, Williams F1 and Lotus F1 among many others
Race Car Conceptual Engineer
E-mail: scalabroni@yahoo.com
Twitter: @ScalabroniE:

Timoteo Briet

Aerodynamic and CFD engineer, Mathematician, Cosmologist, Online Course CFD, Aero and CFD professor
E-mail: racecarsengineering@gmail.com
Twitter: @TimoteoBriet:
https://www.linkedin.com/in/timoteobriet

Nacho Suárez

PhD. Electronics Engineer
Vehicle Dynamics, Virtual 7-Post Rig, Simulation, Autonomous Vehicles, Embedded Systems, Control
E-mail: nachosuamar@gmail.com
https://www.linkedin.com/in/nachosuarezphd/

Collaborator:

Marwan Attar (marwan.attar0@gmail.com), student of our Online CFD Course.

ZONAS PARA OPTIMIZAR EN UN F1; ANÁLISIS AERO-CFD

De Timoteo Briet
con la colaboración de Enrique Scalabroni y Nacho Suárez

Introducción:

En este Artículo, pretendemos conocer qué zonas de un Fórmula 1 son susceptibles de mejorar; no se trata por tanto, de mejorar el coche completo, sino en principio, de mejorar zona a zona; para ello, analizaremos las fuerzas que se producen en cada parte del coche.

Estas fuerzas, básicamente son 2:

- Downforce (o fuerza aerodinámica descendente; la ascendente se suele denominar "lift").
- Drag (o fuerza aerodinámica de arrastre en sentido longitudinal).

Una de las cosas que debemos tener claras a la hora de optimizar un vehículo cualquiera, es saber que una baja presión no indica necesariamente una producción de downforce, o lo que es lo mismo: una alta presión no indica que se esté produciendo una gran resistencia aerodinámica; dependiendo la orientación de una superficie, tenga baja o alta presión, puede producir downforce, lift o más o menos drag.

Analicemos las diferentes partes en que hemos dividido nuestro F1; aquí algunas imágenes generales del coche a estudiar:

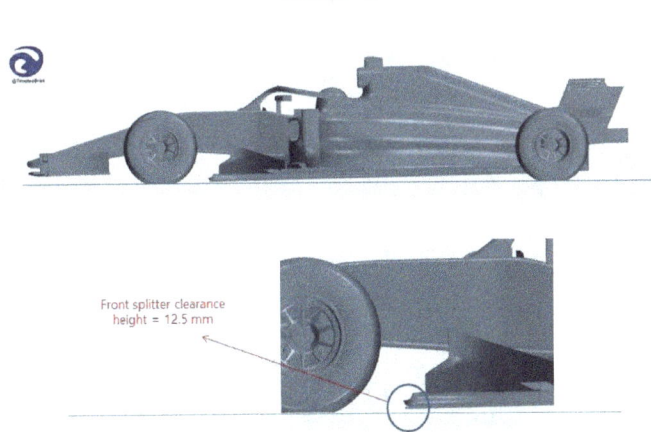

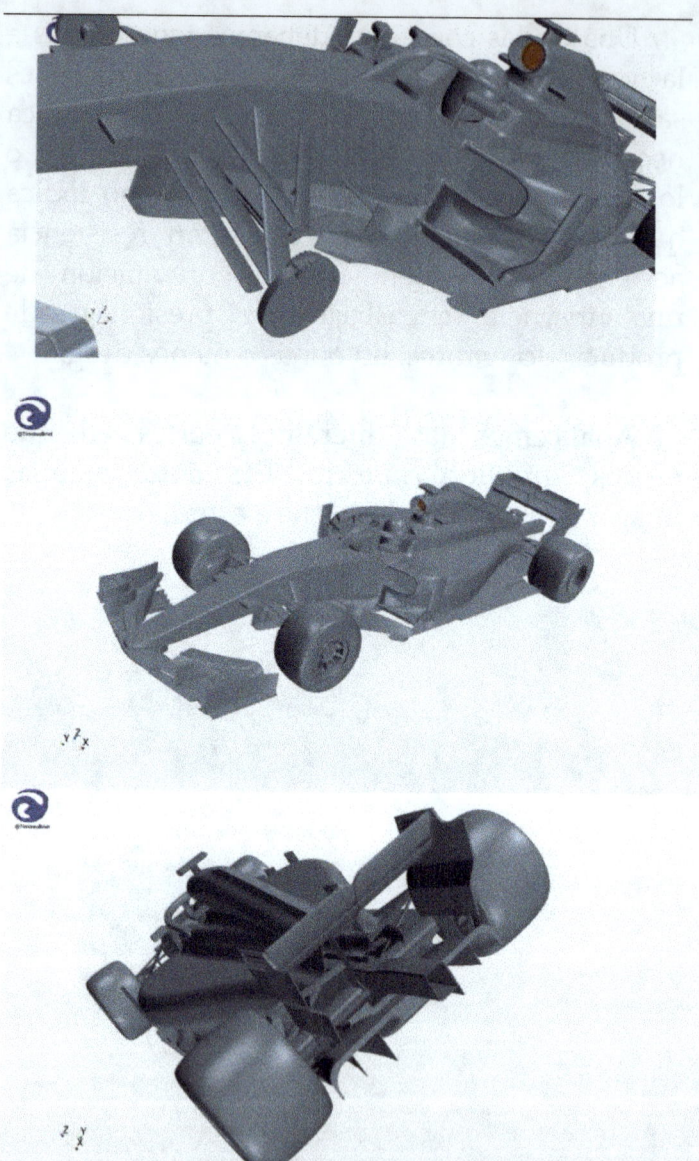

Rake angle = 1.0°
(frontal view)

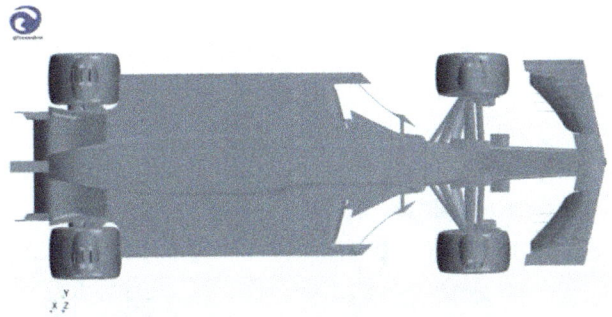

Para determinar las zonas que producen drag y downforce se representan, mediante colores, los coeficientes Cpx y CpZ, respectivamente.

Cpx, indica el coeficiente de drag de una zona determinada; es un valor que puede ser positivo o negativo; en este caso, cuanto más positivo es, mayor drag producirá esa zona: el color rojo indica alta drag y el color azul indica baja drag. Cpz, indica el coeficiente de downforce de una zona determinada; es un valor que puede ser positivo o negativo; en este caso, si es positivo indica una fuerza ascensional o lift (lo contrario de downforce); si es negativo, indica la generación de downforce; el color azul indica generación de downforce, mientras en color rojo indica producción de lift.

Por otra parte también decir que las barras de colores y sus límites, están definidos para comparar subzonas dentro de zonas, no para cuantificar la downforce o drag de cada una de las zonas estudiadas.

Alerón delantero:

En la siguiente imagen vemos que el tercer flap, es el que produce más resistencia aero; el resto de flaps también producen, pero en menor cantidad; las zonas azules, casi mayoritarias, no producen drag.

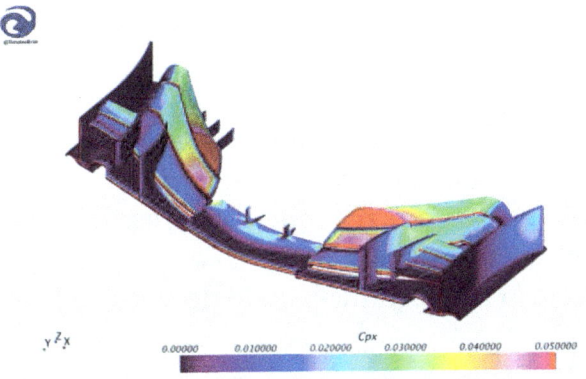

En la zona inferior del Alerón delantero, podemos apreciar algo muy importante y nada intuitivo que seguro desafía mucho conocimiento sobre aero que creamos tener: la parte inferior de los flaps, producen mucha drag; las zonas con color azul y similares, apenas producen drag:

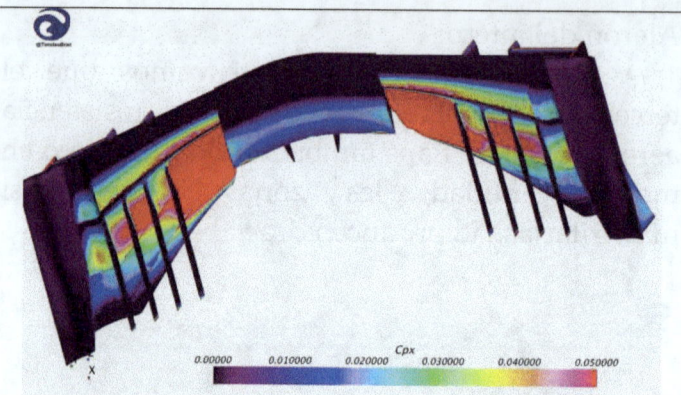

En la siguiente imagen sobre CPz, todas las zonas producen downforce, pero no hay ninguna que produzca mucha carga; estamos hablando de colores amarillos y verdes, indicando la leve producción de downforce:

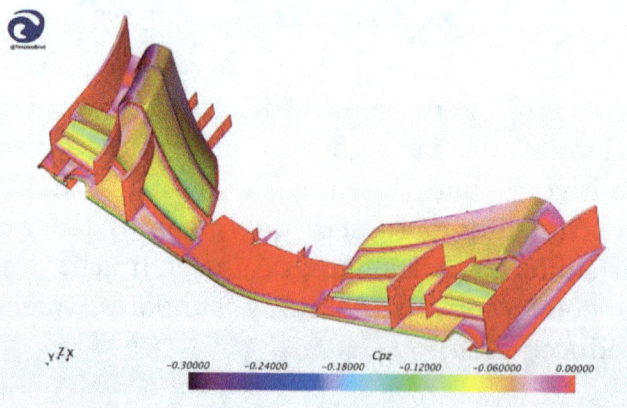

En cambio, en la parte inferior del Alerón delantero, sí podemos ver zonas de mucha generación de downforce; prácticamente toda la parte inferior produce mucha carga, mientras que el resto, si bien también produce, lo hace de una forma más atenuada:

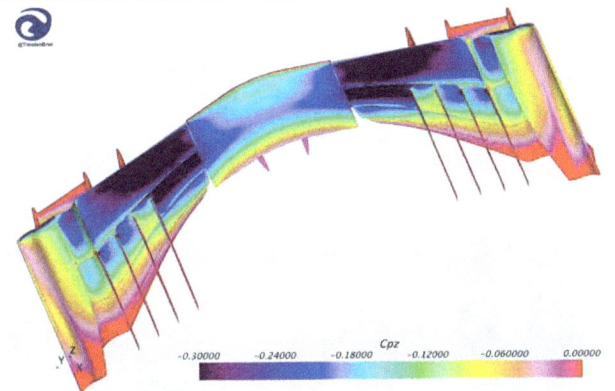

Alerón trasero:

Esta pieza es más simple aerodinámicamente hablando que el Alerón delantero.

Las zonas de generación de resistencia aerodinámica, se circunscriben estrictamente a la pieza superior del alerón trasero; tanto la parte de arriba como la de abajo, generan mucha resistencia; se ve claramente que el color rojo se centra en dicha parte; las otras zonas en la parte de arriba, en amarillo, también produce drag pero en menor medida; las pantallas laterales, en violeta o azul oscuro, no producen prácticamente nada de drag:

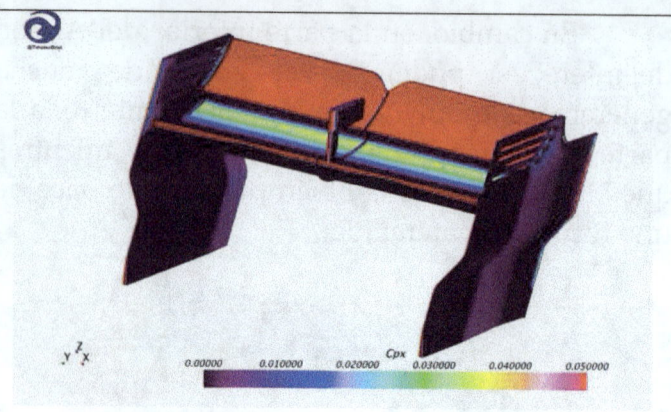

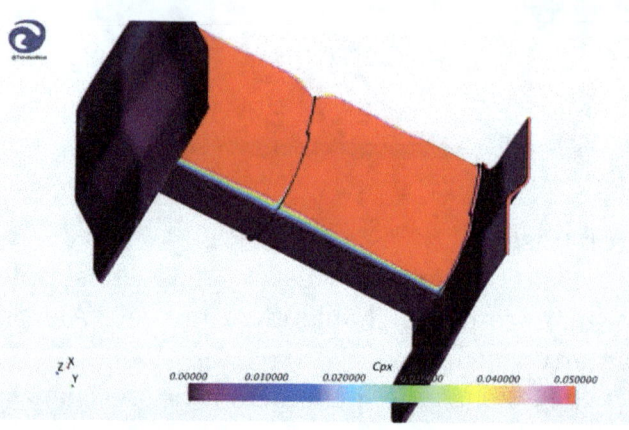

Se ve claramente, que la parte superior del alerón trasero, produce downforce, pero en mucha menor medida que la parte inferior; esto, en ocasiones y a mucha gente, le produce un poco de rareza, ya que el aire impacta bruscamente por la parte superior del alerón trasero, pensando por tanto, que dicha parte, generará más carga que la parte inferior; pero no es así en absoluto; y de hecho, como se puede apreciar, hay gran diferencia entre ambas partes. Es verdad también, que en la parte inferior, el primer alerón es el que más carga produce:

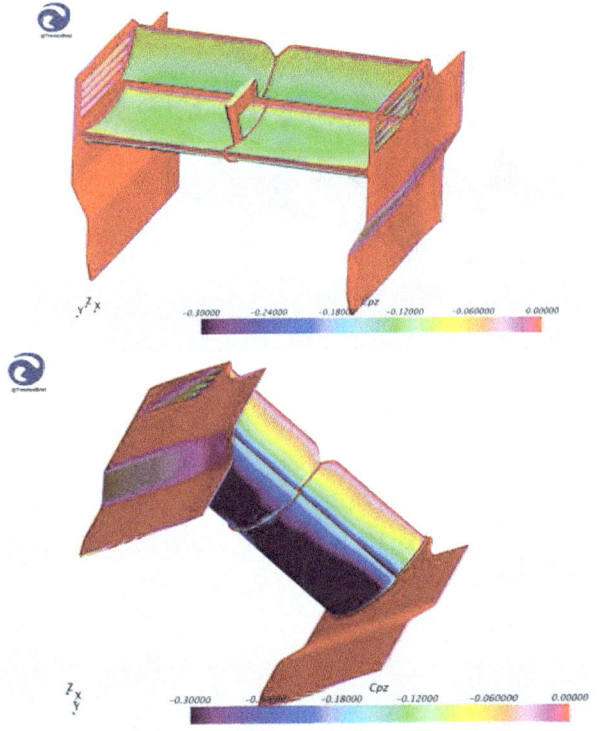

Suelo:

En cuanto a drag, se aprecia perfectamente que la zona inicial del difusor es la que más resistencia aerodinámica produce; el resto del difusor también produce drag, pero en mucha menor medida. Siempre se ha dicho, que la generación de carga por parte del Efecto Suelo, es una generación "gratuita", por cuanto no produce (o apenas nada) drag; podemos ver esto claramente en el color violeta o azul oscuro oscuro de todo el suelo y parte trasera del difusor:

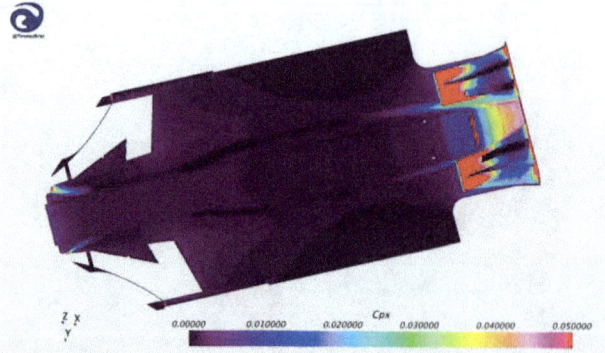

En relación a la downforce, las zonas que más carga producen son la parte frontal de la bandeja y del suelo (debajo de los pontones), así como la parte inicial del difusor; también se aprecia una zona roja en el suelo, que indica la no generación de downforce; ello es debido a que las ruedas traseras se hallan justamente detrás, haciendo de pantalla, produciendo alta presión en dicha zona del suelo y por tanto, no produciendo downforce:

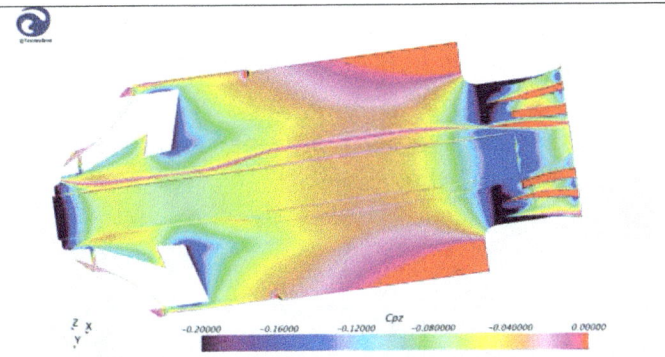

Neumáticos:

Claramente se observa que la parte frontal de las ruedas, es la que más drag produce, cosa lógica:

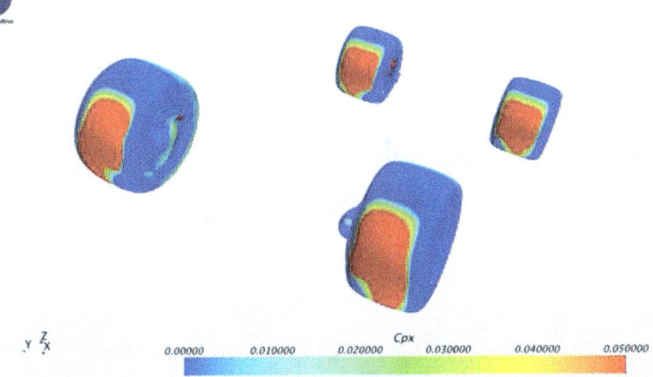

También es muy curioso ver que la parte trasera de las ruedas, producen mucha menos drag que la parte frontal; más curioso: la zona trasera de las ruedas frontales, generan mucha menos drag que las traseras:

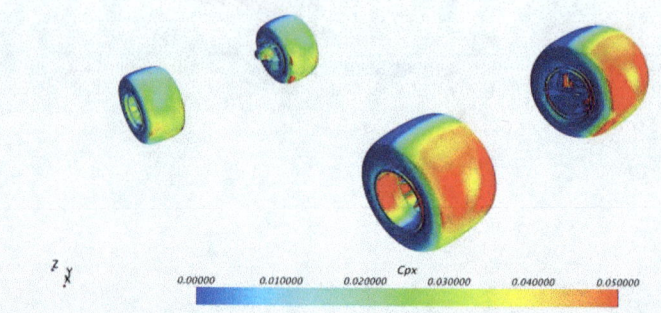

Las zonas que generan carga, se concentran en la parte inferior, cerca de las huellas; son zonas muy pequeñas, no obstante. Se puede apreciar (vista de la parte frontal de las ruedas), una pequeña zona anaranjada, que índice también una pequeña zona de generación de downforce. En la vista trasera de las ruedas (en la segunda imagen), se encuentra también esa zona azul localizada cerca de las huellas:

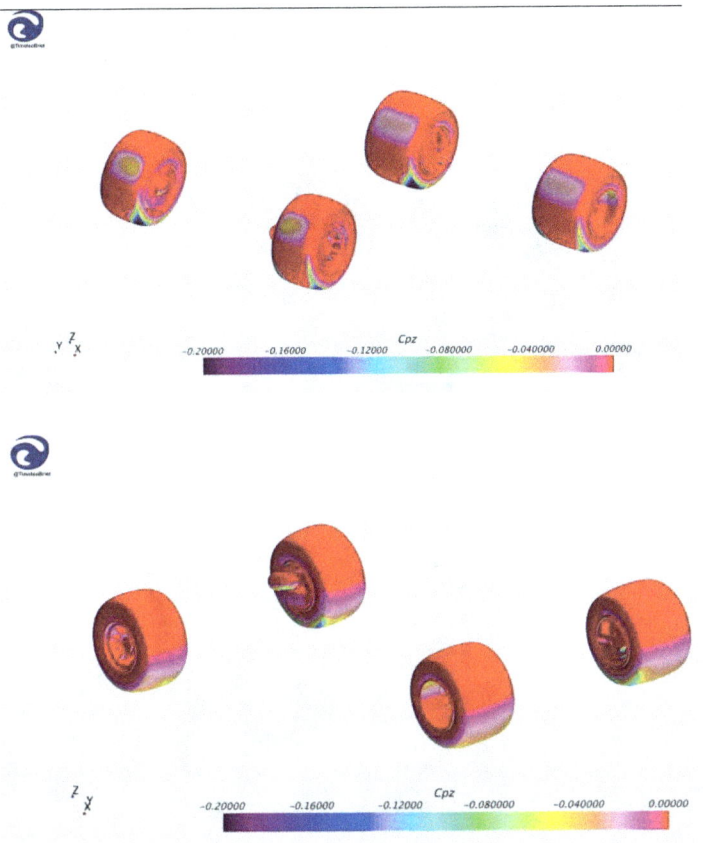

Llantas y sistema frontal de refrigeración de frenos:

Podemos ver la generación de drag por parte de la llanta trasera y del sistema de refrigeración de los frenos delanteros:

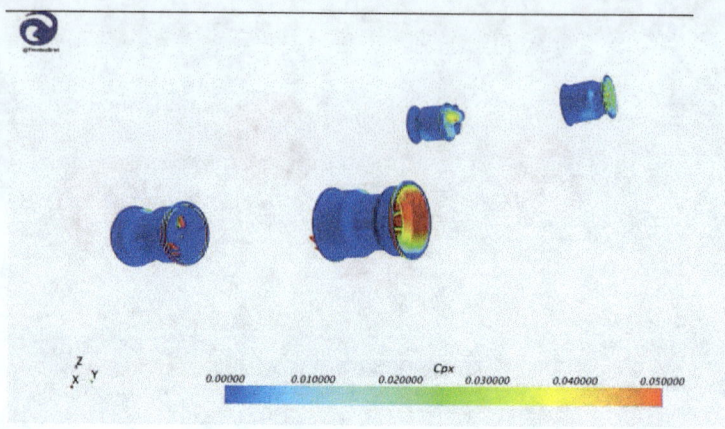

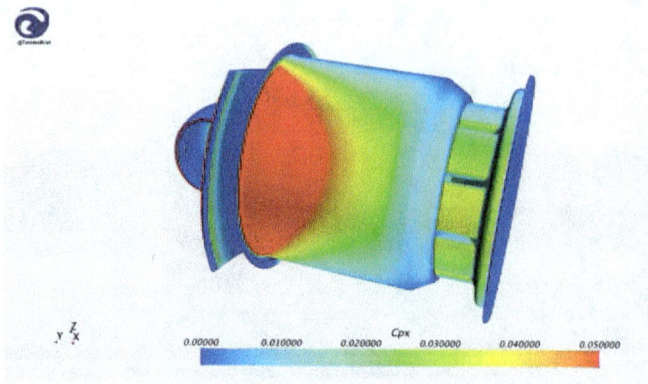

En las llantas, hay pequeñas zonas que producen downforce, al igual que en el sistema de refrigeración de los frenos delanteros; todas ellas se concentran en la parte superior:

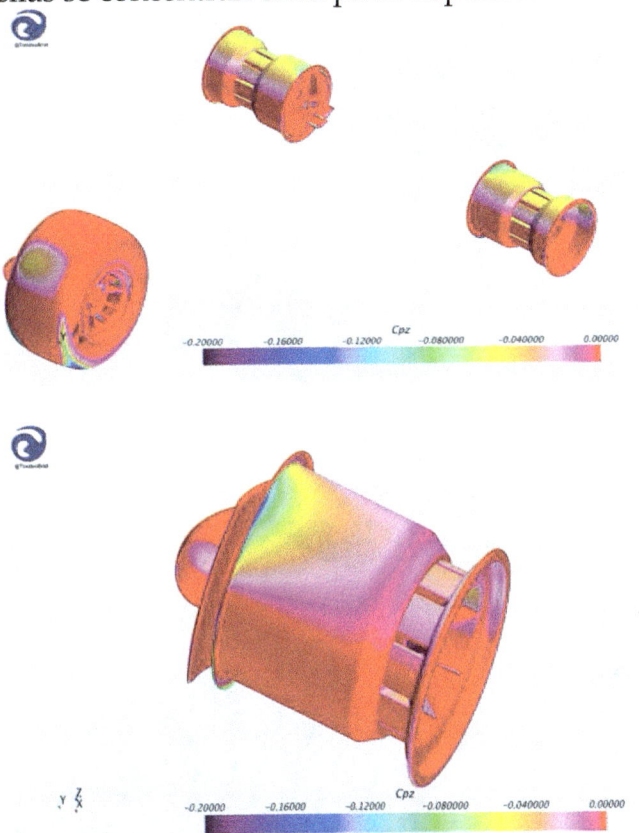

Carrocería:

Las zonas generadoras de drag, se sitúan en las zonas marcadas en rojo:

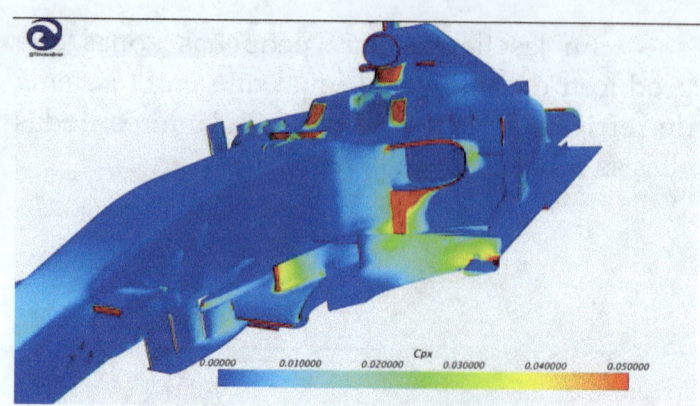

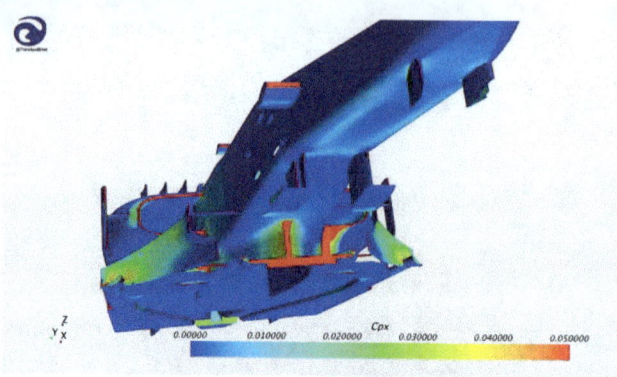

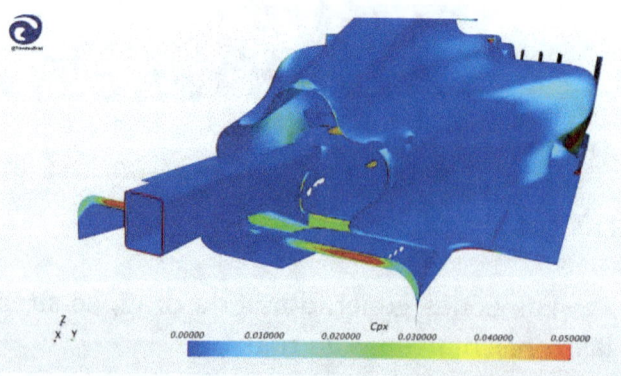

Es una magnífica forma de localizar estas zonas, con el fin de optimizarlas; en este caso, intentar reducir su resistencia.

En cuanto a la generación de carga:

Podemos ya por nosotros mismos, hacernos una idea de las zonas donde se produce carga aerodinámica, sin más que observar la barra de colores y compararlo con las 2 imágenes anteriores; notar por ejemplo entre otras muchas cosas, que hay zonas que producen carga pero también drag. Notar también que en el interior de los pontones, hay una zona en la parte inicial, donde se produce downforce.

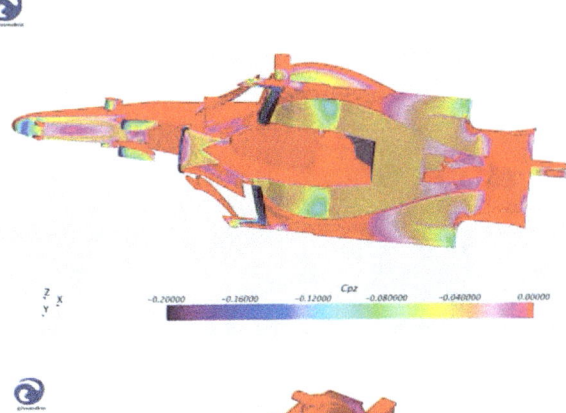

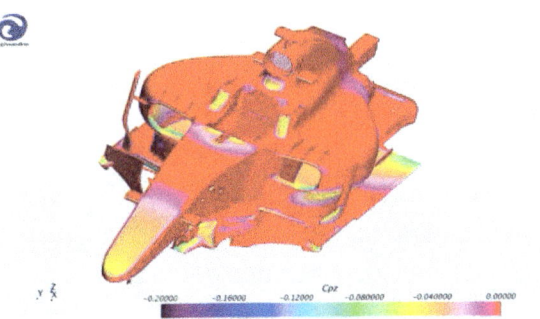

➔ Recordar, que además de representar y localizar por tanto, las zonas con generación de carga y de drag,

Casco:

Por último, decir que este tipo de análisis también se puede aplicar, lógicamente, a elementos pequeños; en este caso: el casco del piloto. En cuanto a la generación de drag:

Sólo hay una zona pequeña, en la parte frontal, donde se genera drag; notar algo importante: las zonas marcadas en azul, amarillo y verde, generan drag "al revés"; es decir: el casco es succionado hacia adelante; es verdad que están en el límite de generación de drag hacia adelante….

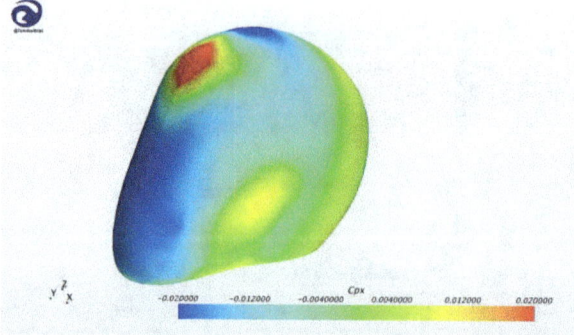

Por último, en relación a la generación de downforce:

Las zonas azules, son generadoras de downforce, mientras que las zonas en rojo, generan lift, es decir: fuerza hacia arriba.

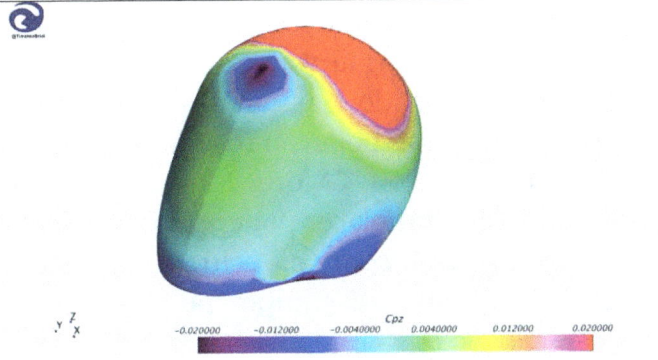

ANÁLISIS AERO – CFD DE UN COCHE FÓRMULA STUDENT / FSAE - 1

Timoteo Briet
Enrique Scalabroni
Nacho Suárez

1) Objetivos:

Este es el primero de Artículos de una serie aplicados a coches tipo Fórmula Student o Fsae o similar; los objetivos principales son 2:

-   Describir el proceso CFD para obtener el balance aerodinámico del coche.
-   Analizar otros valores aerodinámicos en el coche estudiado, a partir de simulaciones CFD.

2) Origen del CAD analizado; proceso de mallado y valores numéricos a calcular; resultados:

Queremos agradecer desde aquí, a Marwan Attar (marwan.attar0@gmail.com), quien ha realizado las acciones pertinentes y necesarias para poder trabajar el coche de 2019 del equipo "More Modena Racing" (Figura 1); es un placer poder trabajar sobre él y aportar ideas sobre su estudio en general, que cualquier otro equipo, puede aplicar a sus vehículos.

Figura 1: Logo del equipo More Modena Racing.
Podemos apreciar el CAD del coche en las siguientes imágenes (Figura 2):

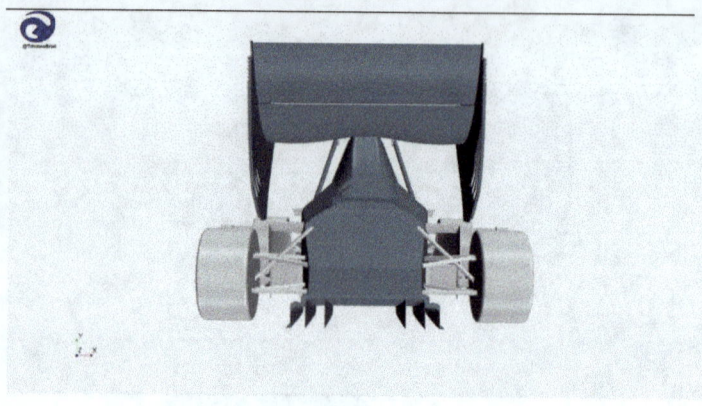

Figura 2: Diferentes imágenes CAD del modelo.

➔ Se trata de un modelo CAD sencillo, el cual, vamos a utilizar para este PRIMER ARTÍCULO; esta es la razón por la cual las ruedas son macizas y los radiadores también son inexistentes. El próximos Artículos, incorporaremos ruedas "reales" con sistema de Refrigeración de Frenos, analizando la transferencia de calor entre el aire y el disco o sistema de frenado, rotación de llantas (unsteady o transitorio), gases de escape (composición química), radiadores, transferencia de calor en escapes, etc....

De igual forma, podemos apreciar el coche ya construido y siendo ensayado en los tests correspondientes (Figura 3):

Figura 3: Modelo "real": More Modena Racing.

La fase anterior al mallado del modelo, consiste en tener perfectamente definido en CAD, el coche; ello significa que los datos numéricos que obtendremos de un análisis CFD, no corresponderán al modelo "real", sino al modelo CAD; por esta razón, el modelo CAD ha de ser lo más parecido posible al modelo físico.

Para obtener un buen modelo CAD completo del coche, y poder hacer pruebas como cambiar un front wing por otro, o un rear wing, o añadir una pequeña pieza, o quitar otra, el coche, geométricamente ha de estar compuesto por piezas ensambladas en el programa CFD; el programa CFD ha de ser capaz de hacer justamente lo dicho anteriormente: sumar piezas o unirlas, restarlas, quitar o añadir piezas, etc…. Cuando esto se ha realizado correctamente, nos encontramos ante el proceso de mallado, que es el siguiente paso.

Al igual que antes se ha comentado que el CAD ha de ser lo más parecido al coche real por cuanto los resultados numéricos que se obtengan corresponderán al CAD, es más: los resultados numéricos serán del CAD cuyas superficies están formadas por la malla que seamos capaces de crear; es decir: el coche se ha convertido en una malla, que ha de ser por tanto, lo más similar al CAD o al coche. De ahí la importancia que tiene realizar una buena malla.

A groso modo, queremos comentar algunos datos importantes acerca el mallado del coche. Ya comentaremos en otros Artículos los tamaños y más, sobre transferencia de calor y rotación frente al tiempo (rotación de la llanta, por ejemplo):

En cuanto a tamaños de malla, normalmente se trabaja con 2 tamaños: tamaño mínimo y tamaño objetivo; el tamaño mínimo ha de ser pequeño, para que la malla se acople a las pequeñas superficies o pequeñas curvaturas; este tamaño puede ser perfectamente 0.5 mm. El tamaño objetivo, es el tamaño al que tenderá en tamaño "general" de la malla; puede oscilar entre 1 cm y 3.5 cm (Figura 4):

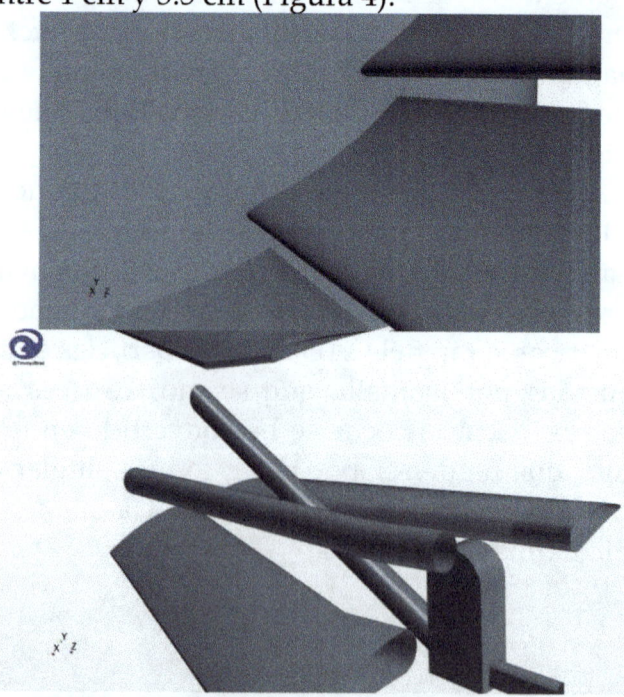

Otro aspecto muy importante a tener en cuenta, es una buena definición de capa límite; se trata de una fina capa alrededor de cualquier superficie del coche, en la cual pasan las cosas más importantes, desde un punto de vista aerodinámico, que definen la aerodinámica global del coche; estos efectos son causados por la viscosidad del aire; el espesor de la capa límite, depende de la velocidad del aire en cada punto, por lo tanto, la capa límite debería tener espesores distintos dependiendo de la zona a implementarla; para hacer esto, se puede hacer una primera simulación CFD (aproximada por tanto), para obtener esos valores de velocidad, y en función de ellos, calcular los espesores en cada punto; y así sucesivamente, haciendo cada vez más, la capa límite más y más correcta; la verdad es que normalmente no se hace esto, a costa de quizás más tiempo de cálculo y mayor peso de la malla; lo que se hace es poner una capa límite igual para todo el coche; el espesor de la capa límite "global" puede ser de 8 mm, con unas 15 capas; existe el llamado factor de Stretching (proporción de espesores entre capas) que puede rondar el valor de 1.4 (Figura 5):

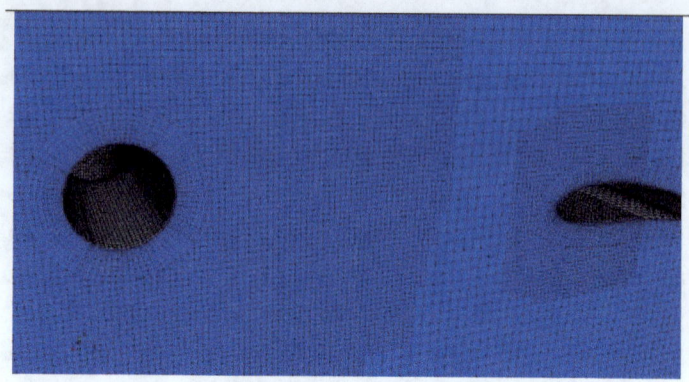

Figura 5: *Capa límite*.

Para saber si se ha colocado una malla en la capa límite correcta, existe un valor numérico denominado "Y+" que indica la idoneidad de la capa creada; valores menores de 25 son muy correctos (Figura 6):

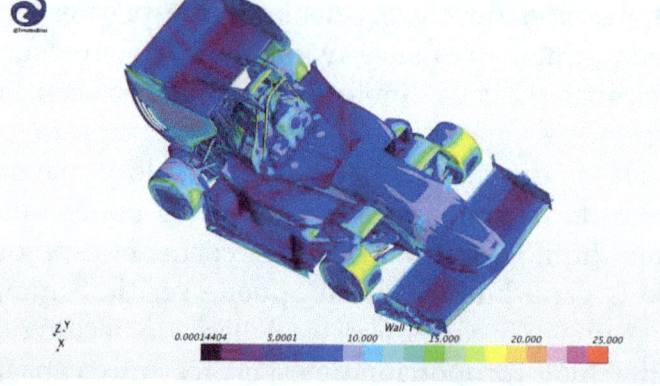

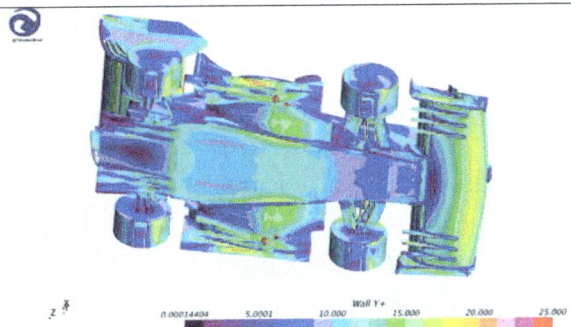

Figura 6: Valores de Y+ de la *Capa límite*.

En nuestro caso, valores entre 25 y 20, apenas hay, lo cual es muy bueno para obtener resultados "reales" con los que poder trabajar.

En cuanto al refinamiento de malla, o lo que lo es lo mismo: hacer la malla más fina y precisa, allí donde se necesita o allí donde pasan las cosas más importantes desde un punto de vista aerodinámico, hay 2 formas de hacerlo:

Colocando volúmenes de control en las zonas deseadas haciendo que en dichas zonas la malla sea más pequeña.

Este método es muy bueno, por supuesto, pero tiene el inconveniente de que entre volumen de control y volumen de control, existen saltos de tamaño de malla, que hay que suavizar, y ello no es fácil; cuanta más diferencia exista entre tamaños, peor es la convergencia de la solución.

O haciendo la malla más pequeña en todo el problema; ello conlleva, claro, mucha más cantidad de malla, pero no hay saltos de malla entre diferentes zonas; ello implica más precisión y sensibilidad de las soluciones.

Los datos generales de las simulaciones que hemos realizado en este primer Artículo, son:
- Velocidad de 50 km/h.
- Rotación de ruedas.
- Suelo rodante.

3) Balance Aerodinámico:

Para calcular el Balance Aerodinámico, se procederá de 2 modos (son prácticamente iguales y se obtienen exactamente los mismos resultados); siempre suponiendo que el eje de coordenadas está situado en el eje frontal del vehículo:
- Calcular Momento de Pitch / Wheelbase = Downforce en el eje trasero; sabiendo esta fuerza y la downforce total, se calcula el balance frontal; es decir: el porcentaje de downforce total que actúa sobre el eje frontal.
- Momento de Pitch / Downforce total = Distancia del eje frontal al centro de presión; sabiendo este punto y el Wheel base, se calcula el balance frontal.

Son 2 métodos "iguales" que permiten conocer el balance del coche.

En este caso y según nuestro análisis CFD, el balance frontal es de un 36.5%. Ello significa, que el 36.5% de la downforce total, actúa sobre el eje frontal.

Veamos la distribución de presiones debajo del coche (Figura 7), para ver de esta forma, las zonas que producen más downforce; de esta forma, podemos conocer las zonas que "trabajan mal"; esta es una de las claves en todo desarrollo aero de un vehículo: conocer las zonas que son mejorables:

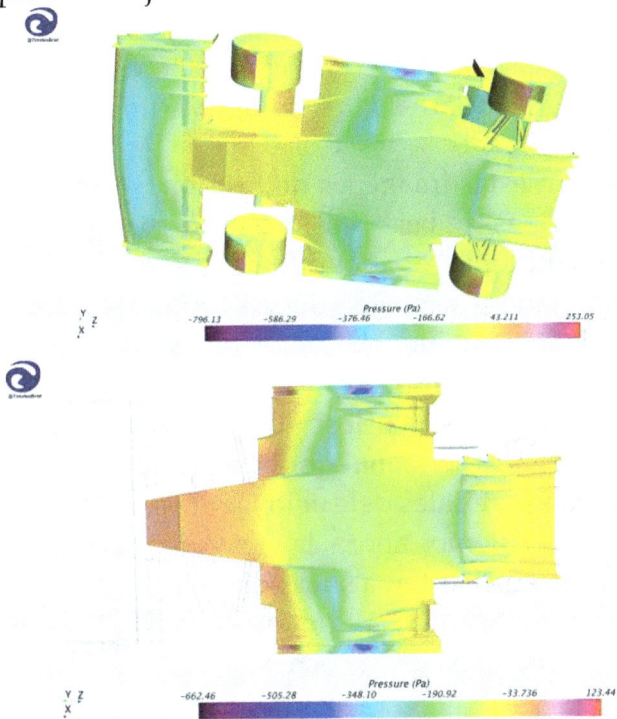

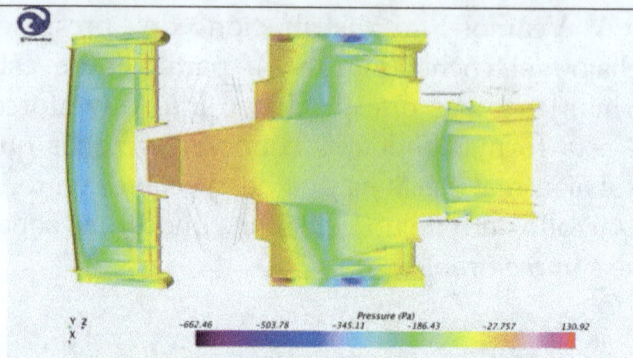

Figura 7: *Presiones sobre la parte inferior del coche.*

En estas imágenes de presión, se aprecian las zonas de baja presión (no necesariamente generadoras de downforce....); en azul, se señalan las zonas de depresión: parte inferior del front wing, y zonas de transición del suelo hacia el difusor y cambio de curvatura. También es curioso (y muy positivo), observar ese zona de baja presión en los canales laterales del suelo (azul oscuro).

Calculemos ahora, la curva de presión que existe en el sistema suelo-difusor del coche (Figura 8), en la sección marcada en blanco en la parte inferior del coche:

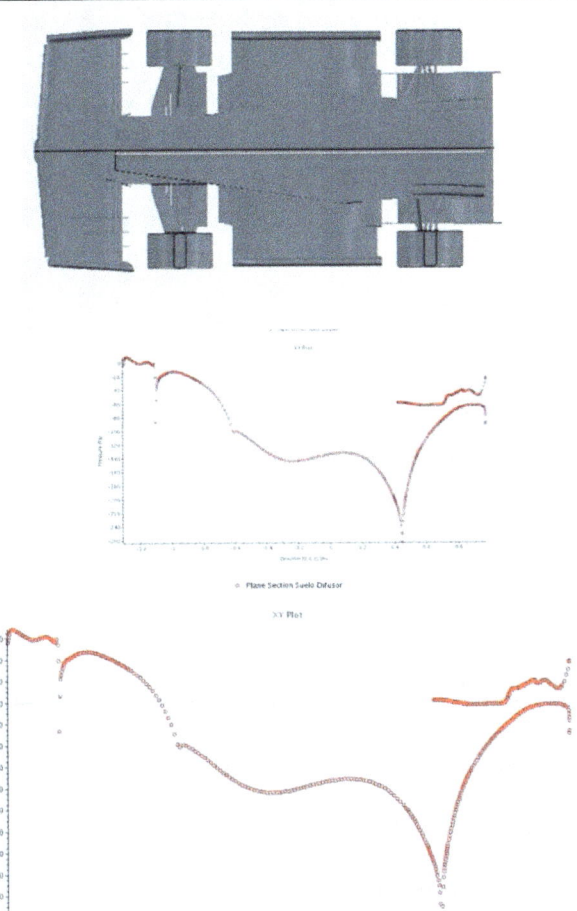

Figura 8: Sección y *Curva de presión en el suelo-difusor*.

Se aprecia perfectamente la zona de transición entre suelo y difusor, donde se produce la bajada brusca de presión que succiona el aire que pasa por debajo del suelo, hacia atrás. Esta es la función del difusor: ayudar al suelo a que funcione bien y produzca la mayor cantidad de downforce posible; de hecho, el difusor no produce demasiada carga; podemos representar en este caso, las cargas (downforce) que realizan cada parte del coche (Figura 9) al igual que las resistencias de cada parte (Figura 10) (los valores de cada tabla, corresponden a Newtons y se refieren a medio coche):

```
Unite.Carena.Faces                    4.980901e+01  -1.415557e-01   4.966746e+01
Unite.cofano.Faces                    7.876097e-01   5.875236e-02   7.963620e-01
Unite.flap anteriorev1.0.Faces       -3.729836e+00  -2.350038e-02  -3.753236e+00
Unite.fondo_2020.Faces                5.114407e+01  -1.163041e-02   5.114291e+01
Unite.front wing.Faces                6.324124e+01  -7.608874e-02   6.316515e+01
Unite.Monoscocca.Faces               -8.933287e+00  -1.110275e-01  -9.044314e+00
Unite.mainhoop.Faces                 -7.456562e-02  -1.195250e-04  -7.468516e-02
Unite.nose.Faces                     -1.757397e+00  -1.465082e-02  -1.772048e+00
Unite.Pilota.Faces                    3.277064e-01   1.333992e-02   3.410463e-01
Unite.rear susp.Faces                -2.061218e-01  -1.676800e-02  -2.228998e-01
Unite.rear wing.Faces                 7.859842e+01  -3.2355666e-01  7.827486e+01
Unite.ruota_rib_ant.Faces            -1.199851e+00  -2.297281e-03  -1.223824e+00
Unite.ruota_rib_post.Faces           -2.531798e+00  -5.922502e-02  -2.591023e+00
Unite.supporto ala post.Faces         3.905087e-01   1.146014e-02   4.019688e-01
Unite.T_Flap_2020.Faces              -1.591200e+00   1.207571e-03  -1.589992e+00
Unite.T_Flap_support.Faces           -4.028215e-02  -1.049541e-03  -4.133169e-02
------------------------------       ------------   ------------   ------------
Totals:                               2.441842e+02  -7.079177e-01   2.434763e+02

Monitor value: 243.47629287186874
```

Los valores de la columna de la derecha, indican downforce cuando son positivos; se ve claramente, que el suelo-difusor, el front wing y el rear wing, son los máximos productores de downforce.

Figura 9: *Downforce de cada parte del coche.*

```
Unite.Carena.Faces              9.725844e+00   1.072008e+00   1.079755e+01
Unite.cofano.Faces              3.456187e-01   1.725460e-01   5.181647e-01
Unite.flap anteriorev1.0.Faces -4.500298e-01   1.058867e-01  -3.441429e-01
Unite.fondo_2020.Faces          9.751862e-01   6.990355e-01   1.674222e+00
Unite.front wing.Faces          9.764008e+00   9.473770e-01   1.071139e+01
Unite.Monoscocca.Faces          1.650712e+01   4.391343e-01   1.694625e+01
Unite.mainhoop.Faces            5.737899e-01   4.701107e-01   6.208010e-01
Unite.nose.Faces                2.516932e+00   7.783000e-02   2.594762e+00
Unite.Pilota.Faces             -1.148147e-01   1.710564e-02  -9.770907e-02
Unite.rear susp.Faces           3.512470e-01   2.634561e-02   3.775926e-01
Unite.rear wing.Faces           2.798826e+01   1.223512e+00   2.921177e+01
Unite.ruota_rib_ant.Faces       2.818338e+00   1.320404e-01   2.950378e+00
Unite.ruota_rib_post.Faces      1.055382e+00   5.252742e-02   1.107910e+00
Unite.supporto ala post.Faces  -1.142736e-01   5.828383e-02  -5.598972e-02
Unite.T_Flap_2020.Faces         1.990525e-01   2.269587e-02   2.217464e-01
Unite.T_Flap_support.Faces     -4.778287e-03   4.651830e-03  -1.264574e-04
--------------------------      ------------   ------------   ------------
Totals:                         7.213658e+01   5.097991e+00   7.723458e+01

Monitor value: 77.23457521055136
```

Figura 10: *Drag de cada parte del coche.*

En este caso, la eficiencia (Downforce / drag) es de aproximadamente 3.

Siguiendo con una de las premisas esenciales en todo diseño "preliminar" de un vehículo, y en particular en relación a la detección y análisis de las zonas que non trabajan como deben trabajar, una de las representaciones más ilustrativas que se pueden realizar, es aquella en la que podemos observar qué zonas del coche producen Downforce; es decir: aquellas zonas que producen fuerza en sentido vertical hacia la pista (Figura 11); se ha representado sólo los colores correspondientes a valores de coeficientes de downforce entre 0 y -1; ello significa que se representan sólo los colores de las zonas que producen downforce.

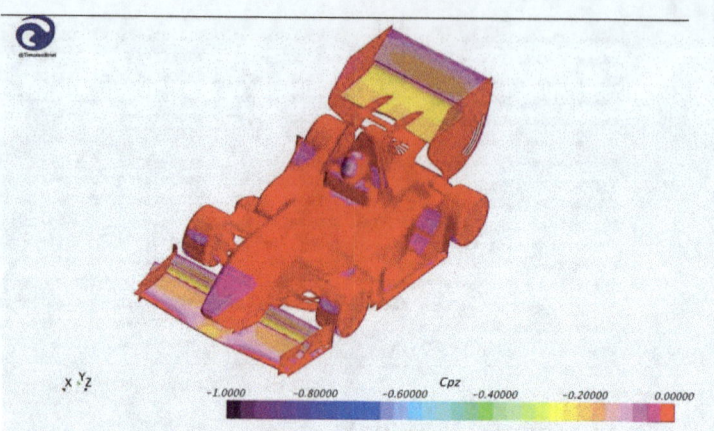

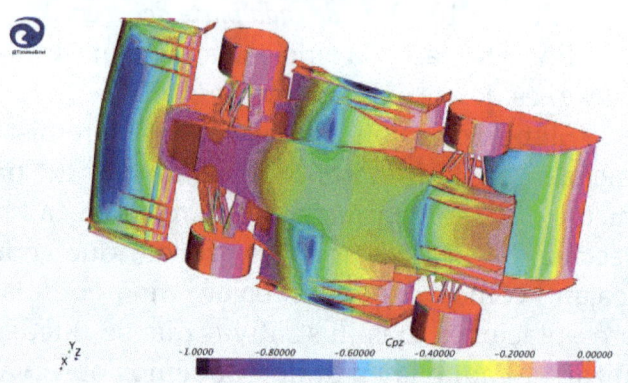

Figura 11: *Zonas del coche donde se produce
downforce.*

➔ Imaginad, por ejemplo, visualizar las zonas que
PRODUCEN DRAG; sería un método increíble de
localizarlas para poder optimizarlas; estamos
pensando un Artículo aplicado a un F1, justamente
con este tipo de representaciones, con el objetivo de
conocer y localizar dichas zonas posibles de
mejorar. O lo que es lo mismo: en las imágenes
anteriores, se pueden representar las zonas del
coche que PRODUCEN LIFT….

En el rear wing, podemos apreciar que la parte inferior, no trabaja bien completamente, pues hay como una especie de "U"; debería ser toda constante de color (Figura 12):

Figura 12: *Rear wing: zonas de creación de downforce.*

Otra forma d optimizar las zonas del vehículo, es conocer las zonas que producen drag "reversible"; es decir: Drag que se genera pero que es posible eliminar; por ejemplo, podemos optimizar las partes frontales de los Strakes que hay debajo del front wing; hay color rojo (alta presión) en la parte frontal de cada uno de ellos (Figura 13):

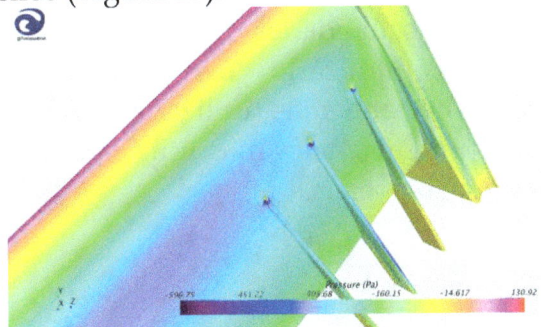

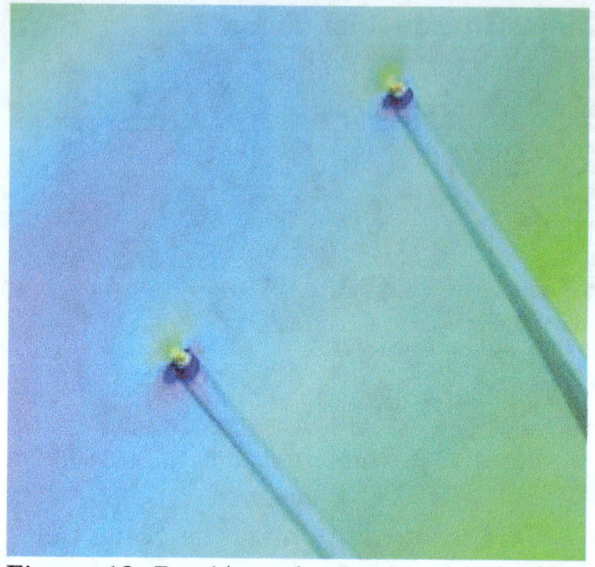

Figura 13: *Presión en los Strakes del front wing y alrededores.*

Conclusiones:

Con este Primer Artículo de una serie de trabajos aplica dos a este coche, pretendemos dar a conocer las posibilidades que tiene un estudio CFD para optimizarlo aerodinámicamente: métodos y herramientas utilizadas, valores a calcular, representaciones más útiles, proceso de mallado y cálculo, etc....